Charlie Brown's Second Super Book of Questions and Answers

about the earth and space...from plants to planets!

Charlie Brown's Second Super Book of Questions and Answers

about the earth and space...from plants to planets!

Based on the Charles M. Schulz Characters

Random House New York

Editor: Hedda Nussbaum
Art Director: Eleanor Ehrhardt
Designer: Terry Flanagan
Layout: Charlotte Staub

Special thanks to:

David A. Dundee
Intern Astronomer and Scientific Assistant
Hayden Planetarium
American Museum of Natural History, New York

Carlos R. Dunn
Meteorologist in Charge
Weather Service Forecast Office, Atlanta
formerly
Chief, Scientific Services Division
National Weather Service Eastern Region

Dr. George Harlow
Assistant Curator
Department of Mineral Sciences
American Museum of Natural History, New York

Jane Herrman
Supervisor of Education
Morris Arboretum, Philadelphia
formerly
Traveling Instructor
Brooklyn Botanic Garden, New York

Photograph and Illustration Credits: Lockheed Missiles and Space Company, 140; National Aeronautics and Space Administration, 99, 103, 106, 111, 112, 113, 114, 115, 121, 124, 125, 131, 132, 135, 144; Tass from Sovfoto, 123.

Library of Congress Cataloging in Publication Data
Main entry under title: Charlie Brown's second super book of questions and answers about the earth and space . . . from plants to planets! SUMMARY: Charlie Brown and the rest of the Peanuts gang help present scientific facts about plants, geology, weather, climate, astronomy, and space travel. 1. Natural history—Miscellanea—Juvenile literature. 2. Science—Miscellanea—Juvenile literature. [1. Natural history—Miscellanea. 2. Science—Miscellanea. 3. Questions and answers] I. Title: Charlie Brown's second super book of questions and answers about the earth and space. QH48.C49 500.1 77-7191
ISBN 0-394-83491-7   ISBN 0-394-93491-1 lib. bdg.

Manufactured in the United States of America

4 5 6 7 8 9

Introduction

Have you ever wondered why the ocean is salty? Or what makes bread get moldy? Or why you can see your breath on a cold day? Or who owns outer space? Or why you can't ever see the back of the moon? If you have, this is the book for you. It will give you the answers to all these questions and to many, many others—about plants, rocks, oceans, rain, snow, tornadoes, earthquakes, stars, planets, rockets, astronauts, and more.

Once again the whole Peanuts gang is here to help out with the answers. So join good old Charlie Brown and start asking some questions!

Contents

What is a plant?

Anything that is alive and isn't an animal is a plant. Unlike animals, most plants stay in one place. They don't walk, swim, or fly. Most plants have green leaves. In them is the chemical chlorophyll (KLAWR-uh-fill), which gives the leaves their green color. A few plants with chlorophyll don't have green leaves. They have red, purple, or brown leaves instead. Any plant that has chlorophyll is called a "green plant" even if no green shows on its leaves. Green plants make their own food. Other plants don't. Instead, they take their food from animals or other plants. Non-green plants may be completely brown, white, or even red.

How many kinds of plants are there?

There are about 350,000 different kinds of plants on earth. They come in all sizes. Some are so tiny that you can see them only under a microscope. Others are so large that they tower hundreds of feet above the ground. In fact, the tallest living thing is a plant—the giant redwood tree. It can grow as high as 300 feet (90 meters).

Plants have many different shapes, too. A blade of grass is long and skinny. A palm tree has large leaves and a long trunk. A cabbage is round and leafy. A mushroom is shaped like an umbrella. A cactus is usually narrow with sharp spines.

How long have plants been on earth?

Plants have been on earth for more than three billion (3,000,000,000) years and possibly as long as four billion (4,000,000,000) years. The first plants were tiny water plants—the kind you can see only under a microscope. They were on earth many millions of years before dinosaurs. In fact, they were here long before any animals.

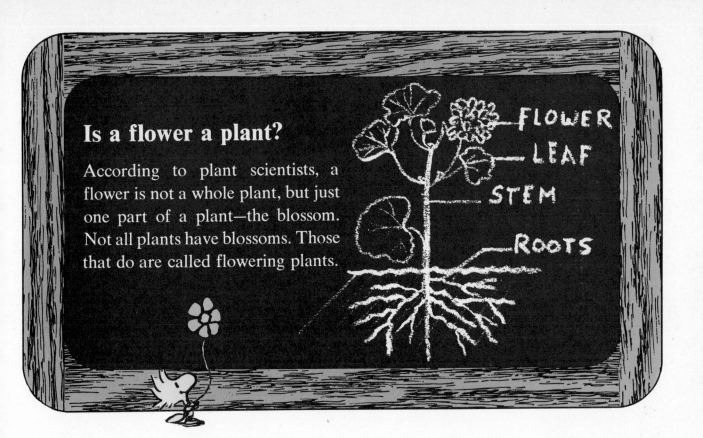

Is a flower a plant?

According to plant scientists, a flower is not a whole plant, but just one part of a plant—the blossom. Not all plants have blossoms. Those that do are called flowering plants.

FLOWER
LEAF
STEM
ROOTS

Why do green plants need leaves, roots, flowers, stems?

A green plant makes its own food. Most of the food is made in its leaves.

Roots hold a plant firmly in the ground so that it does not fall over or blow away. The roots also take water and minerals out of the soil. The plant needs these things to live. Sometimes roots store some of the food that the leaves make.

The flowers are the parts of the plant where seeds can form. The seeds will someday become new plants.

The stems hold up the leaves and flowers. They have tubes in them that carry liquids up and down the plant. Some of the tubes bring water mixed with minerals from the roots to the leaves and flowers. Other tubes carry liquid food away from the leaves to the rest of the plant. We call the two liquids sap.

The largest flowering plant growing on earth today is a Chinese wisteria (wih-STEER-ee-uh). It has branches 500 feet (150 meters) long, and it weighs as much as 50 elephants!

3

How do leaves make food?

The leaves of a green plant are like little food factories. Inside them is the chemical chlorophyll, which makes the leaves green. The leaves need the chlorophyll in order to make food. When the sun shines on the chlorophyll, each leaf factory goes to work.

The factory uses two things to make food. It uses water that has come up from the soil, through the roots and stems. It uses a gas called carbon dioxide (die-OCK-side) that has come from the air, through tiny openings in the leaf. From the water and carbon dioxide it makes sugar, which is the plant's food. At the same time, the factory makes a gas called oxygen (OCK-suh-jin). Most of this oxygen goes into the air.

The food factory works only when the sun is shining. When the sun goes down, the factory stops. Without sunlight (or an electric plant-light) green plants cannot make food and they die.

Why do leaves of house plants turn toward the window?

If you let a green plant stand on your window sill, its leaves will turn toward the window. This may happen after a few hours or a few days. If you turn the plant around, the leaves will again move and face the window. Why? Because light is coming in the window, and green plants need all the light they can get to make food. The leaves are reaching for the light.

Outdoors, most green plants have bright light all around them, so their leaves don't turn. But they, too, would turn if light reached them from only one direction.

NEST FOR SALE

A pine needle is really a leaf!

4

What is a carrot?

A carrot is the root of a carrot plant. This root is filled with stored food. When the leaves of a plant make food, the plant uses some of it for energy to grow. Extra food is stored in the plant's roots, stems, fruit, seeds, and even in its leaves. A carrot plant stores most of its extra food in its root. When you eat a carrot, you are eating this stored food.

Carrots are not the only roots you may eat. Beets are roots. So are sweet potatoes. Like carrots, they are filled with stored food. Celery stalks are stems that have food stored in them. So are asparagus stalks. Lettuce, cabbage, and spinach are leaves with a little food stored in them. Apples, peaches, and grapes are the fruits of plants. Peas, lima beans, and corn kernels are all seeds. They, too, hold stored food.

What is a fruit?

We usually think of a fruit as a sweet and juicy food. But plant scientists think of fruits in another way. To them a fruit is the seed or seeds of any plant together with the fleshy parts around them. For example, string beans, eggplants, and tomatoes all have seeds inside them. So to a plant scientist they are fruits, although most of us would call them vegetables. Apples, oranges, cherries, and bananas are fruits both to us and to the plant scientists.

What do plant scientists mean by the word "vegetables"? They never use that word. They talk about roots, stems, and leaves—but not about vegetables.

Do all plants make their own food?

No. Only green plants—plants that have chlorophyll—can make their own food. All other plants cannot. A plant that doesn't make food is called a fungus. More than one of these plants are called fungi (FUN-jI). Some fungi take their food from dead wood or from soil that has rotting plants in it. Others take their food from living plants and sometimes from living animals. Some fungi even live on people. A rash called athlete's foot is caused by fungi living on a person's skin.

The fungi you probably know the best are mushrooms. These usually grow in areas where there is very little sun. Because mushrooms don't make their own food, they don't need sunlight.

What's the difference between a mushroom and a toadstool?

Toadstools are poisonous mushrooms. Toadstools got their name because people used to think that poisonous toads sat on them. Unless you are an expert, it is almost impossible to tell a poisonous mushroom from a good one. If you just want to touch mushrooms, they all are safe. But if you want to eat them, you'd better buy your mushrooms at the supermarket.

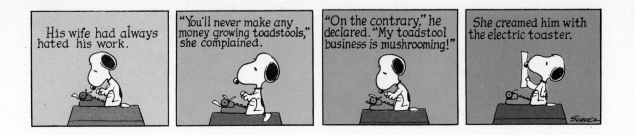

What makes bread get moldy?

All around us in the air are tiny black and green specks called spores. They are too small for us to see without a microscope. But they are there all the same. These spores are like seeds for certain plants called molds. Molds are fungi and do not make their own food. They live off some of the foods we like to eat. Bread is one of them. When bread is moldy, you know that bread-mold spores have landed there. The spores have grown into fuzzy-looking mold plants which are eating the bread.

The medicine penicillin comes from a mold that is very much like bread mold!

Why do lakes and rivers sometimes turn green and slimy?

Dirty water from toilets is often dumped into lakes and rivers. And fertilizers, which make farmers' crops grow better, are often washed by rain into lakes and rivers, too. In these same waters live tiny green plants called algae (AL-jee). Both the fertilizers and the human wastes make the algae grow faster. They grow so fast that they may cover the whole top of a lake or a large part of a river. They make the water look green and slimy. Any fish living in this slimy water will die.

What is the crust you sometimes see on rocks?

The gray or colored crusts that you see on rocks are plants called lichens (LIKE-inz). You can find them growing not only on rocks, but also on tree bark and even on sand. You can find lichens in hot deserts and on cold high mountains where no other plants can grow.

How can lichens live in these harsh places? Lichens are made up of two kinds of tiny plants living together—algae and fungi. Lichens don't need wet soil to get water because both the algae and fungi can take water out of the air. The algae, which are green plants, also made food for the lichens. No one is sure what else the fungi do. But when living together as lichens, the algae and fungi can survive in places where they could not live alone.

9

Venus's-flytrap

Do any plants eat animals?

Yes, there are plants that eat insects and one that can eat small birds and mice! Three of the animal-eating plants are the pitcher plant, the sundew, and Venus's-flytrap.

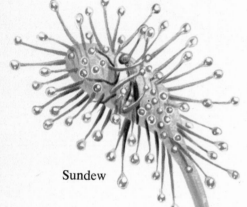

Sundew

The pitcher plant has leaves that are shaped like pitchers or vases. The pitchers have a sweet smell that attracts insects. At the bottom of each pitcher is water. If an insect falls inside, it drowns. The plant then digests the insect in much the same way that your body digests food. One kind of pitcher plant has pitchers so large that small birds and mice sometimes fall inside and are digested.

The sundew has leaves covered with many hairs. Each hair has a drop of sticky liquid at its tip. In the sunlight, these drops of liquid shine like dew. They also smell sweet. Insects are attracted to the plant. When an insect lands on a leaf, it gets caught in the sticky drops. The hairs bend over the insect and hold it down. Then the plant digests it.

Venus's-flytrap is well named. Each leaf can fold along the center and close up like a trap. On each leaf are little hairs. When an insect lands on these hairs, the two parts of the leaf close, trapping the insect inside. After the insect is digested, the leaf opens again.

Even though the pitcher plant, the sundew, and Venus's-flytrap eat insects, they are all green plants and make their own food as well. However, certain minerals are missing from the soil in which they live. They get these minerals from the insects they eat.

10

Pitcher plant

Why do bees fly around flowers?

Bees go to flowers to get nectar and pollen. They use the nectar, a sweet liquid found inside many flowers, to make honey. They take home the pollen, a powder in the flowers, to eat and to feed to their young.

SOMEHOW I FEEL RIDICULOUS EXPLAINING THE BIRDS AND BEES TO WOODSTOCK!

Do bees do anything for flowers?

Yes, bees help flowers make seeds. Most flowers can't make seeds unless they get some pollen from another flower of the same kind. Bees bring this other pollen to the flowers when the bees fly around gathering food. Here's how it works:

Bees go to only one kind of flower at a time. When a bee lands on a flower, it brushes against tiny spikes called stamens (STAY-minz). On the stamens is pollen. Some pollen brushes off onto the bee's furry body. Then, when the bee lands on another flower, some of this pollen falls off its body.

Inside flowers is another spike called a pistil. When the pollen from one flower lands on the pistil of another flower of the same kind, we say the plant has been pollinated. Now seeds can start to grow. Seeds will not grow unless the flower is pollinated. From the seeds will come new plants someday.

Bees are not the only things that pollinate flowers. Sometimes the wind carries pollen from one plant to another. Often butterflies, moths, wasps, flies, beetles, birds, and even bats do the job. They go to the flowers looking for nectar. The flowers attract the creatures with special smells or colors.

11

Do all flowers smell sweet?

No. Only flowers that attract bees, butterflies, moths, and some other insects have a sweet smell. Flowers that attract birds do not. Birds have a poor sense of smell, but they are attracted to bright colors. "Bird" flowers are usually bright red or orange. They also hold a lot of nectar. Only birds that like nectar pollinate flowers.

Flowers that are pollinated by flies have a smell, but not a sweet one. These "fly" flowers smell like rotting meat! "Bat" flowers also smell bad to people, but bats like the smell. Flowers that are pollinated by the wind have no smell at all. They don't have any nectar either, and they have very dull colors. These flowers don't need to attract anyone. They are often small and hard to notice.

Why does the dandelion turn white and fluffy?

The dandelion flower turns white and fluffy so that the wind can carry away its seeds. After a dandelion flower has been pollinated, seeds begin to grow inside it. When the seeds are ripe, the yellow petals fall off the flower. Fluffy white hairs sprout at the top of each seed. The wind catches the fluff and carries the seed away from the plant. After a while, the seed lands on the ground. If the seed gets pushed into the ground by an animal or rain, the seed will grow into a new plant.

Because dandelion seeds are spread by the wind, new dandelions can grow in many places. If all the dandelion seeds simply dropped to the ground right under the plant, the seeds would be too close together. When they sprouted, the new plants would not have room to grow well.

Are all seeds spread by the wind?

No. All seeds are spread in some way, but not all by the wind. The wind spreads seeds that are tiny or fluffy and can easily float in the air. It also spreads seeds that have "wings," such as maple seeds.

Water sometimes spreads these winged seeds, too. Any seeds that are able to float can be spread by water. Such seeds may travel long distances before they reach land again. Animals also spread seeds. Some seeds have sharp hooks that get stuck to an animal's fur and fall off later. Burrs, or "stickers," are seeds of this kind. Some birds and other animals eat fruits. They may carry off the fruits and then drop the seeds far away.

One kind of daisy is pollinated by a snail!

FANTASTIC!

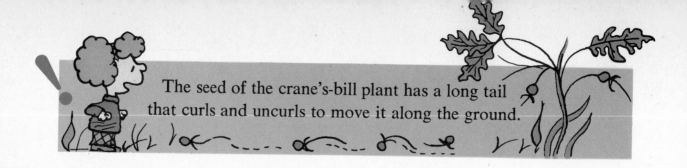

The seed of the crane's-bill plant has a long tail that curls and uncurls to move it along the ground.

How does a seed become a new plant?

If you have ever eaten a sunflower seed, you have had to crack open the hard shell outside to get to the soft part inside. Some of this soft part is the beginning of a new sunflower plant. It is called an embryo (EM-bree-oh). The rest of the soft part is stored food. Inside every seed is an embryo with stored food. The hard shell protects the embryo.

The embryo forms while the seed is still on the plant. Once it is formed, it stops growing for a while. But it has enough stored food around it to give it energy to grow again. It will grow as soon as it is in the right soil with the right temperature and the right amount of water. Then the embryo will sprout roots and a stem. It will grow leaves and start making its own food. It will become a whole plant and make seeds of its own.

What is a pine cone?

When a pine cone first grows, it is a kind of flower. Later on it acts as the tree's fruit because it holds seeds. When the pine cone is still a flower, it is covered with soft scales. After it has been pollinated by the wind, seeds begin to grow inside it. The scales harden and protect the seeds. The seeds may grow inside the hardened cone for several years. But at last they become ready to fall to the ground or to be blown away by the wind.

Other trees that are related to the pine also have cones. Firs, spruces, and cedars are some of these trees.

14

Do all new plants come from seeds?

Plants that have flowers come from seeds. But other plants start in other ways.

Some very tiny plants—the kind you can see only under a microscope—make new plants simply by dividing in half. Each part becomes a whole new plant. Other microscopic plants grow "buds." These buds have nothing to do with flowers. They are whole plants, smaller than their parent, but otherwise just like it. A bud stays attached to the parent until it grows to nearly full size. Then it splits off from the parent plant and lives on its own.

Some algae and fungi grow tiny specks, called spores, instead of seeds. From each spore can come a new plant. Look under the umbrella of a ripe mushroom and you will see the mushroom spores.

Ferns also produce spores. These are inside small brown cases that you can see lined up on the underside of the fern's leaves. When the spores are ripe, the cases burst open and the spores are scattered.

Ferns come out of the ground all curled up, and then they unfold!

Why do people put plant stems in water?

People put flower stems in water to make the flowers last longer. People put leafy stems in water to grow roots. Once roots have grown, the stems can be planted in soil. Then they will grow into full-sized plants. The plants that can grow roots on their stems also have seeds. But seeds may take a few weeks or months to begin growing. Some stems can grow roots in water in just a few days. You can grow roots on the stems of many plants, including begonias, ivy, and coleus (KOE-lee-us) by putting them in water.

Are there any other quick ways to grow plants?

Yes. Sometimes stems with leaves on them will grow if you simply put them in soil and water them well. You can grow a pineapple plant from the top of a pineapple this way. Cut off the leafy top, strip some of the lower leaves off the stem, and stick the stem in a pot of soil. With the right amount of water and sunlight, the stem will grow into a whole new pineapple plant.

Set a sweet potato in a jar of water so that some of the sweet potato sticks out above the jar. Roots will grow from the part under water. Long stems with leaves will grow out of the dry top part.

You can also grow many new potato plants from one white potato. On each potato you will see little dark bumps. These are called eyes. Cut up a potato so that each piece has one eye. If you plant the pieces in soil, a new plant will grow from each eye.

What is a tulip bulb?

A tulip bulb is a special underground part of the tulip plant. On the outside of the bulb is a brown skinlike covering. Inside is the bud of a flower and a short stem with thick leaves packed tightly around it. The leaves hold extra food for the plant so that it can live through the winter and grow again in the spring.

In spring the tulip leaves, stem, and bud come out of the bulb. They push above the ground. The tulip bud grows and opens. After the flower is pollinated, the flower petals fall off the stem, and seeds fall to the ground. But the leaves are still on the plant, and they are still making food. The food is stored in the underground bulb. After the summer, the leaves and stem fall off the plant. Above the ground, the plant cannot be seen. But underground there is still a living plant in the form of the roots and bulb. In the fall a new stem, bud, and leaves form in the bulb. The plant stays underground all winter, using the stored food. When the weather gets warm again, the bulb sends up the new leaves, stem, and bud. The whole story repeats itself year after year.

Although most people plant bulbs to get tulip flowers, you can also grow tulips from their seeds. But if you do, you will have to wait three to seven years before you get any flowers.

What is a weed?

A weed is any plant that grows where people don't want it to grow. For example, when farmers are growing potatoes, they don't want any other plants growing in the same soil. Other plants will take minerals and water from the soil that the farmers want their potatoes to have. So they call any other plant a weed. Farmers spray their potato fields with special chemicals to keep weeds from growing. They dig up any weeds that do sprout there. Plants whose seeds scatter easily are often called weeds because they come up in many unwanted places.

How can you tell how old a tree is?

When a tree is cut down, you can usually see rings on the tree stump. The rings show how long the tree was growing. Twice each year, a tree grows some new wood. The new wood forms a ring around the old wood of the tree trunk. In the spring, the tree grows a light-colored ring. In the summer, it grows a dark-colored ring. So if a tree stump has 24 rings, you know that the tree was 12 years old when it was cut down. If the stump has 200 rings, you know that the tree was 100 years old.

HERE'S THE WORLD FAMOUS BEAGLE SCOUT TEACHING HIS TROOPS HOW TO TELL THE AGE OF A TREE.

Some bristlecone pine trees have lived nearly 5,000 years.

Why do leaves turn colors in the fall?

Leaves have many colors in them—green, red, orange, yellow. But during the spring and summer, there is much more green than any other color. The green comes from the chlorophyll which leaves use in making food. The leaves have so much chlorophyll that you usually can't see the other colors. But in the fall, before cold weather sets in, many trees stop making food. At the same time they stop making chlorophyll. When the chlorophyll disappears, you can see the other colors in the leaves.

Why do leaves fall off the trees in the fall?

During warm weather, the leaves of a tree are always giving off tiny drops of water. At the same time, the tree's roots are taking in more water so that the tree does not dry out. But during cold weather, the ground freezes. The roots cannot get any water. If the leaves kept on giving off water, the tree would dry up and die.

In the fall, a layer of cork grows at the bottom of each leaf stem. It blocks water from flowing into the leaf. The leaf dries up. It is easily taken off the tree by the wind. And so it falls to the ground.

19

Why is a cactus covered with spines?

The thin, sharp spines of a cactus protect it in two ways. First, they keep the cactus from losing a lot of water. Plants are always losing water through their leaves. Plants that have large leaves may lose a few gallons a day. A cactus lives in the desert where there is very little water. If it lost much water, it would quickly die. So, instead of large leaves, the cactus has small spines. They are so thin that little water can escape through them.

The spines also protect the cactus by being sharp. A cactus stores a lot of water in its stem. Desert animals might break open the cactus to get this water, but the sharp spines keep the animals away.

Why does poison ivy make you itch?

The poison ivy plant has an oil on it. This oil can make your skin break out in an itchy rash. If you know what a poison ivy plant looks like, you can keep from touching it.

Poison ivy can grow as a bush or as a vine that climbs on other trees. The plant's leaves are each made up of three leaflets (three leaves on one stem). They are shiny and green in spring and summer. They turn a bright red in the fall.

Poison ivy

Does a four-leaf clover really bring good luck?

You are lucky to find a four-leaf clover because most clover plants have only three leaves on each stem. But there is no way that a four-leaf clover or any other plant can bring you good luck.

At least two people have been lucky enough to find a ten-leaf clover!

What are tree surgeons?

Tree surgeons act as doctors for trees. They take care of trees, keeping them healthy and healing them when they are hurt or sick. Tree surgeons cut back branches that have gotten too long, spray chemicals on insects or fungi living in a tree, paint a protective coating over cuts, and take off dead wood. In a way, tree surgeons are also like dentists because they fill in cavities, or holes, in trees. The holes are made by fungi, insects, or woodpeckers.

Do plants feel pain when they are cut?

No. Animals feel pain because they have nerves inside them that carry messages of pain to their brains. But plants don't have nerves or brains. So they can't feel pain the way we do.

Do plants grow better if you talk to them?

No one knows for sure. Many scientists think that sound can affect plants. Soft, quiet music seems to help them grow better. Loud music with drums seems to kill them. But plants can't understand words, so saying nice things to a plant should not affect it. Still, some scientists claim that plants react to the feelings in people's words—and even in their thoughts. They say that plants grow better if you think nice thoughts and that they "faint" or even die if you think nasty thoughts. But other scientists don't believe that these things are true. Many people have done experiments to find out if they are. So far everyone seems to have come up with a different answer.

Which foods that people eat come from plants?

Cacao tree

Fruits and vegetables, of course, come from plants. But they are not the only foods that do. Coffee comes from the seeds of the coffee plant. Chocolate comes from the seeds of the cacao (kuh-KAY-O) plant. Honey is made by bees from the nectar in flowers. Sugar can be made from two different plants—sugar cane or sugar beets. Sweet maple syrup comes from the sap of the sugar-maple tree. Wine is made from grapes. Meat and milk don't come from plants—but they're from animals that eat plants. In fact, one way or another, all our food comes from plants—except salt and chemical food made in factories.

22

What would happen if all the plants on earth died?

If all the plants on earth died, so would all the animals—including people. We need plants in order to live. When green plants make food, they give off oxygen. This is a gas that all animals must breathe in order to stay alive. Without plants, animals would have no oxygen to breathe and would die.

Animals also depend on plants for their food. All animals eat either plants or plant-eating animals. Without plants there would be almost no food on earth!

IT SAYS HERE THAT "WITHOUT PLANTS THERE WOULD BE ALMOST NO FOOD ON EARTH." ...THAT KIND OF TALK GIVES MY STOMACH A HEADACHE.

FLOATING OUT TO SEA ON A PITCHER'S MOUND... I CAN'T BELIEVE IT!

CHARLIE BROWN'S IN TROUBLE, SNOOPY... WE SHOULD DO SOMETHING...

THAT'S TRUE!

IF HE'S NOT GOING TO BE AROUND TO FEED ME ANY MORE, MAYBE I SHOULD PLANT A GARDEN...

LET'S SEE, I COULD PUT SOME TOMATOES HERE, AND SOME CORN OVER THERE AND MAYBE SOME RADISHES HERE...

23

What things that we use every day come from plants?

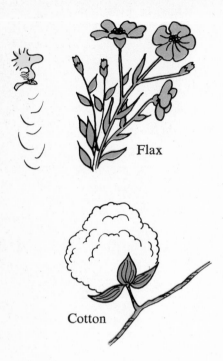

Flax

Cotton

Everything made of wood comes from the trunk of a tree. Houses, fences, furniture, and paper are all made of wood.

From the sap of the rubber tree comes rubber. Tires, shoe heels, rubber bands, rubber balls, diving suits, and many toys are made of rubber.

Cotton cloth is made from the cotton plant. Linen cloth is made from the flax plant. From cotton and linen cloth we make clothes, curtains, towels, and sheets.

Many medicines come from plants. Penicillin comes from a mold. Quinine, used to treat the disease called malaria, comes from the bark of the cinchona (sin-KOE-nuh) tree. Digitalis (dij-ih-TAL-iss), used to treat weak hearts, is made from dried leaves of the foxglove plant.

What is recycled paper?

Recycled paper is new paper made out of old. Most paper is made from the trunk of a tree. Each year people use more and more things made of paper—towels, napkins, plates, cups, books, magazines, newspapers, and stationery. To get all this paper, trees are cut down every day. New trees are often planted in their place, but new trees take a long time to grow. In order to save some trees we need to use less paper or reuse old paper.

Scientists have invented a way to reuse paper. We call this "recycling" the paper. Here is how it is done. Old newspaper is first shredded into little pieces. Then it is put into chemicals that clean it and make it soft and mushy. The mushy paper is mixed with water, flattened, and spread to dry. When it is dry, it is fresh, clean paper—recycled paper.

The Earth

What is the earth made of?

The earth is a great ball of rock. Underneath its grass, soil, oceans, and rivers lie thousands of miles of rock.

If you could dig a hole deep into the earth, here is what you would find. At first you would see hard rock, like the kind you see above ground. The rock would feel cool when you touched it. This rock is part of the crust, or outside, of the earth. As you went deeper, the crust would become hotter and hotter. When you got about 5 miles (8 kilometers) into the earth's rock it would be hot enough to roast you alive! If you could keep digging in spite of this heat, you would reach the part of the earth called the mantle. This starts about 20 miles (32 kilometers) below the ground. Most of the rock here would be hard. But some would be soft and gluey—like very thick molasses. And the temperature would still be rising. The center, or core, of the earth may be as hot as nine thousand degrees Fahrenheit (9,000° F., or 5,000° C.)! A lot of this core is liquid rock.

No one has ever dug far enough to see or feel what the earth is like deep inside. However, scientists have machines that can gather information without ever going below the ground.

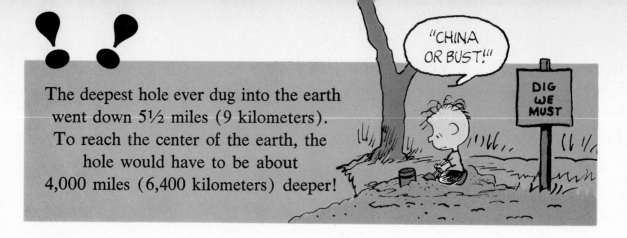

The deepest hole ever dug into the earth went down 5½ miles (9 kilometers). To reach the center of the earth, the hole would have to be about 4,000 miles (6,400 kilometers) deeper!

"CHINA OR BUST!"

DIG WE MUST

Why is the inside of the earth hot?

THE ANSWER IS THE DEVIL MADE IT THAT WAY!

Many scientists believe that billions of years ago the earth was a great ball of dust and gas. Over a long period of time, the bits of dust and gas moved closer and closer together. Finally, they joined to become solid rock. All the movement caused by the shrinking made the earth heat up. In fact, it got so hot that the rock melted into a gluey liquid. After millions of years, the outside of the earth—the crust—cooled off. Because it cooled, it became hard rock again—in the same way that melted chocolate gets hard when it cools in the refrigerator. But the inside of the earth did not cool. It has stayed hot until today because certain minerals in the earth give off a lot of energy and heat. We say that these minerals are radioactive (RAY-dee-oh-AK-tiv).

26

Does hot rock ever come out of the earth?

Yes. Hot, liquid rock called lava comes out of volcanoes. Scientists give the name "volcano" to any crack in the earth's crust from which lava flows.

Scientists are not sure what makes a volcano act up—or erupt. But they think that hot gases inside the earth push lava up from below. The force of these gases may also cause the loud noise that people sometimes hear when a volcano erupts.

The earth may shake when a volcano erupts. Out comes fiery-hot, glowing lava. Steam, ashes, and even solid rocks pour out, too. Once the lava reaches the surface of the earth, it cools and hardens. Often so much lava, rock, and ash come out that a mountain builds up around the crack. Then the whole mountain, with its crack, is called a volcano. A big volcano mountain can take as long as ten thousand years to build up. But one volcano in Mexico grew 200 feet (60 meters) high in a single day.

No volcano keeps erupting all the time. The rest period of a volcano may be just a few minutes, a few months, a few years, or several hundred years. And some volcanoes stop erupting altogether. We say these volcanoes are extinct.

IT SAYS HERE SCIENTISTS ARE NOT SURE WHAT MAKES A VOLCANO ERUPT.... WHICH REMINDS ME A LOT OF MY SISTER.

VERY FUNNY... OH, VERY FUNNY!!!

VOLCANOES ARE LITTER BUGS

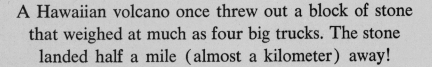

A Hawaiian volcano once threw out a block of stone that weighed at much as four big trucks. The stone landed half a mile (almost a kilometer) away!

Are volcanoes dangerous?

Yes. The hot lava that pours out of a volcano often causes fires. Sometimes so much lava flows that it buries a whole city or a whole island. Also, when a volcano erupts, it often sends out a cloud of smoke filled with poisonous gases. The gases spread over a large area and kill everyone who breathes them in.

Volcanoes are most dangerous on islands. Because islands are surrounded by water, people have a hard time escaping.

Do volcanoes do any good?

Yes. The inside of a volcano is very, very hot. It heats everything around it, too. At several places in the world, people have learned to use this heat.

Much of the earth's water is underground. The underground water around an active volcano—one that erupts from time to time—is very hot. Wells have been drilled near such volcanoes. Pipes have been put down the wells, deep enough to reach the boiling-hot underground water. Up through the pipes has come steam given off by the boiling water. The steam is used to turn machines that make electricity.

The greatest gift of volcanoes is their help in making rich soil. All soil is made mostly of crumbled rock. Soil made from volcanic rock has in it many of the minerals that help plants to grow well. The land around a volcano is the best farmland in the world. However, farmers must live with the knowledge that the volcano may erupt again at any time.

What is a hot spring?

Some parts of the earth have natural hot or warm water under the ground. If this water comes up above the ground through a crack, people call it a hot spring.

Some hot springs are hot because their water comes from far down in the earth where there is great heat. But the water from most hot springs starts closer to the top of the earth. This water is hot because it is near a volcano. The volcano may be an active one, or it may be one that is extinct.

What is Old Faithful?

Old Faithful is a geyser (GUY-zur) in Yellowstone National Park, Wyoming. A geyser is a special kind of hot spring. Its water gets so hot underground that it boils and explodes into steam. The geyser spurts the hot water and steam into the air from an opening in the ground. The water shoots up like a fountain for a while, and then dies down. The water in some geysers rises only a couple of inches. In others, it shoots as high as a ten-story building. Some geysers spurt only once every few years. Old Faithful got its name by shooting water faithfully about once every hour. It shoots the water as high as a seven-story building. Old Faithful's water is hot because about a million years ago small volcanoes existed in the Yellowstone area.

29

What is an earthquake?

Any snapping or breaking of the earth's crust is called an earthquake. The snapping makes the earth shake, or quake. If you were to snap a plastic ruler in two, you would see the two halves shaking for a few seconds after the snap. That is what happens to the earth during an earthquake. But in the earth, the shaking lasts longer than a few seconds.

Forces inside the earth are always squeezing and straining the rock of the earth's crust. Scientists aren't sure why this happens. Usually the forces cause the rock to bend, but not snap. So there is no earthquake. But sometimes the forces are so great that they make the rock snap. If you are close to where the rock has snapped, you feel the earth shiver. You feel the earthquake.

Can an earthquake change the surface of the earth?

Yes. A big earthquake can break off part of a mountain, which then tumbles down onto the land below it. An earthquake can tear open the ground. It can shove huge blocks of land around. Any of these things can happen in just a few minutes.

Big earthquakes cause a lot of damage. Buildings fall down. Gas pipes burst and start fires. Whole cities sometimes start burning. Water pipes break, so there is no water to put out the fires. Many people are killed by the falling buildings or the fire. Luckily, most earthquakes are small. They do very little damage.

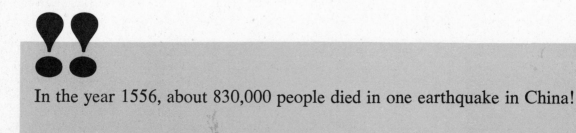

In the year 1556, about 830,000 people died in one earthquake in China!

How can you protect yourself from earthquakes?

The best way to protect yourself from earthquakes is to stay away from the areas where they happen. Earthquakes do not happen in all parts of the world. They happen in special areas called earthquake belts. One runs along the western coasts of North and South America.

   If you *do* live in an earthquake belt, it helps to live in a well-built house with a steel frame. The house should be built on solid rock—not on soft clay. If an earthquake does begin, don't panic. Don't rush out into the street. People have been trampled to death during earthquakes by crowds of panicky people in the streets.

   Today, scientists have machines that measure the movements in the earth. They can sometimes tell when an earthquake is coming, so that people can get out of an area before the quake starts.

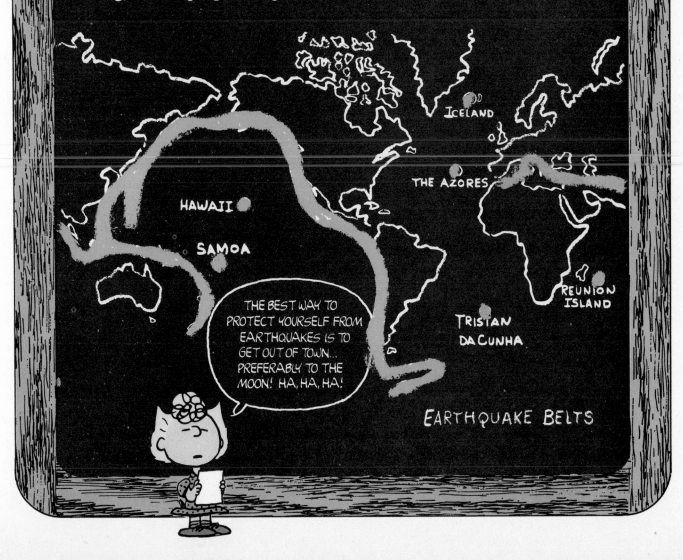

What is a fossil?

A fossil is the remains of a plant or an animal that lived millions of years ago. Some fossils are shells and bones of animals that have turned to stone. Some are leaf prints left in rock. Others are tracks of animals, such as dinosaurs, that hardened into stone. These tracks look like footsteps made in cement.

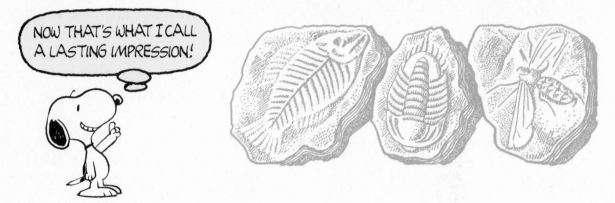

How were the mountains made?

Many mountains were made from rock that was pushed up from the bottom of the ocean. Scientists know this because fossils of ancient sea animals are buried in the tops of the highest mountains.

Mud and sand—called sediment (SED-uh-mint)—are always being carried by rivers from the land down into the ocean. Sediment that was carried to the ocean many millions of years ago came to rest in low places on the ocean floor. The skeletons of sea animals became mixed with the sediment. For hundreds of thousands of years, sediment piled up in layers on the ocean floor. The sand, mud, and skeletons got packed and squeezed together into solid rock. After many more thousands of years, forces inside the earth squeezed the rock into folds—the way you can squeeze the skin on the back of your hand into folds with your thumb and another finger. Finally, these forces pushed the folded rock upward to make many of the mountains we see today.

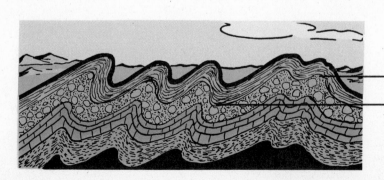

Rock is pushed up into mountains.

Layers are squeezed into folds.

How many different kinds of rock are there?

There are three groups of rock. All the rocks you can ever find belong to one of these three groups.

The first group is called igneous (IG-nee-us) rock. It started out deep under the ground. At one time it was so hot that it was a gluey liquid. Most igneous rock cooled and hardened underneath the earth. But some of the liquid—lava—broke through to the earth's surface. It flowed out from volcanoes, and then hardened. One type of igneous rock, granite, is often used for statues and buildings because it is very strong.

The second group of rock was made from sand, mud, or clay that rivers washed down from the land into the sea. It was packed down on the ocean floor in layers. Later, much of this rock rose again to make mountains. This kind of rock is called sedimentary (sed-uh-MEN-tuh-ree) rock. Cement is made from a sedimentary rock called limestone.

The third kind of rock is one that was once either igneous or sedimentary rock. But for millions of years it was bent, folded, twisted, squeezed, and heated by forces in the earth. And so it was changed into a different kind of rock. This kind is called metamorphic (met-uh-MORE-fik) rock, which means "rock that has been changed." The "lead" in a pencil is really graphite, which comes from a metamorphic rock.

Most rock is very hard and stiff. But itacolumite (it-uh-KOL-yuh-mite) is so flexible, you can bend it with your two hands!

What are rocks made of?

All rocks are made of one or more minerals. If you look closely at most rocks, you will see speckles in them. These speckles are minerals. You can see some of them shine if you hold the rocks in bright light. The size, shape, and pattern of the minerals can help you figure out what kind of rocks you're looking at.

There are about 2,500 known minerals in the world. They have different colors, feel different to the touch, and have different strengths. But all of them are made of pieces called crystals. Minerals are found only in nature. They are never man-made. Some minerals you may know are quartz, copper, gold, and diamonds.

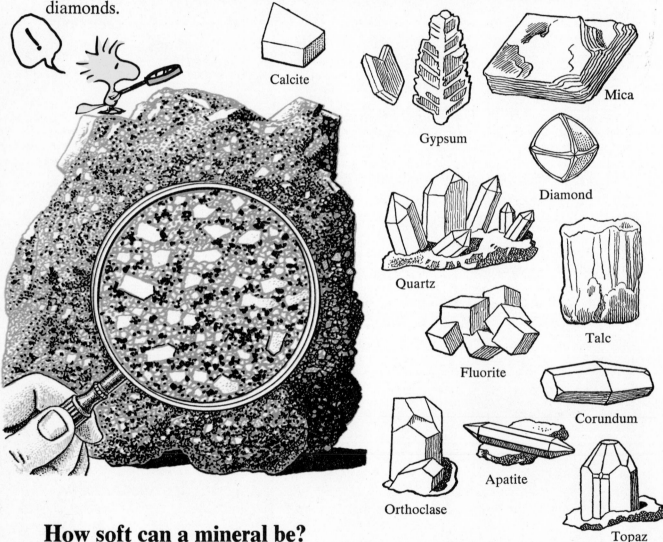

Calcite

Gypsum

Mica

Diamond

Quartz

Talc

Fluorite

Corundum

Apatite

Orthoclase

Topaz

How soft can a mineral be?

A mineral can be soft enough to be scratched by your fingernail. This means that your fingernail is harder than the mineral. Talc is one of these very soft minerals. It is so soft, in fact, that baby powder is made from it! Most minerals, however, are harder—some a lot harder and some just a little.

What is the hardest mineral?

A diamond is the hardest of all minerals. The only thing that can scratch a diamond is another diamond. Because diamonds are so rare and hard and beautiful, they are very valuable and are used to make rings and other jewelry.

How do people get minerals out of the earth?

People get useful minerals such as tin and copper out of the ground by mining. Most minerals are not easy to get out of the earth. Miners tear them out with picks and shovels. Sometimes miners have to drill the minerals out of the earth, or blast them out with explosives such as dynamite.

Is all mining done underground?

No. When people mine a mineral that lies near the top of the earth, they don't have to dig underground. To get at the mineral, miners can just strip off a thin layer of soil with machines, such as bulldozers. This kind of mining is easier and faster than mining far underneath the ground. No long holes and tunnels have to be dug. However, miners have ruined the land in some places by mining from the top. They have not put the soil back where it belongs so that plants can grow there again.

GOOD GRIEF!

I FOUND THREE PENNIES HERE TODAY. I BET THERE'S A COPPER MINE BURIED UNDER THE GROUND.

Is coal made of minerals?

No. Coal is made of the remains of plants that lived many millions of years ago. At that time much of the earth's land was flat and swampy. In the swamps grew huge forests of ferns, mosses, and large trees. As the plants died, they fell into the swamps and began to rot. New plants grew over them. Then they died and new plants grew over them. This happened over and over. Finally, the top layers of dead plants packed and squeezed those at the bottom into a spongy material called peat.

After millions of more years, the swampy land sank in many areas. Water ran into the low places. Mud and sand were washed into the water and covered the peat. The weight of the mud and sand packed and squeezed the peat even more. The peat became buried inside the earth, where it was very hot. The heat, together with the heavy weight, finally turned the peat into hard, black coal.

36

Did cave men make caves?

No, cave men didn't make caves. They found the caves and lived in them because they had no other homes.

Most really big caves are made of a soft rock called limestone. For thousands of years, rain water kept dripping down through tiny holes in this rock. Very slowly, the water wore away the rock, forming hollow caves.

At first, caves are filled with water. But over a long period, most of the water runs out.

What are the stone "icicles" found inside many caves?

Water runs along the ceiling of a limestone cave, picking up bits of a mineral called calcite (KAL-site) that is in the limestone. Some of the water goes up into the air—evaporates. But the calcite does not. So it is left on the ceiling. As more water evaporates, more of the calcite is left on the ceiling. It begins to form a bump. Slowly, more and more water runs over the bump and drips downward. The bump grows down from the ceiling like an icicle. It is called a stalactite (stuh-LAK-tite).

Water that drips to the floor of the cave also evaporates. The calcite from the water builds up on the floor. The calcite forms what looks like an upside-down icicle. This is called a stalagmite (stuh-LAG-mite). Sometimes a stalagmite grows up until it joins a stalactite that is growing down. Together they form a long column from the floor to the ceiling.

FANTASTIC!

What is soil?

Soil is the dark-brown covering over most land. It can be a few inches or a few feet thick. Some people call soil "dirt."

Soil is made mostly of tiny bits of rock of different sizes. It also has in it tiny pieces of plants and animals that have died. Mixed in with soil, too, are tiny living things such as bacteria (back-TEER-ee-uh)—germs so small you need a microscope to see them. Soil also has water and air in it.

How was soil formed?

Billions of years ago, when the earth was young, there was no soil. Only water and rock lay on the surface of the earth. Rain and wind began to beat against the rock. Swift rivers and ocean waves pounded at the rocks. Slowly they wore the rock down. Water seeped into cracks in the rock. In cold weather, the water froze. Frozen water—ice—takes up more space than liquid water. So the ice pushed against both sides of a crack. It split the rock into stones. Rain and rivers washed the stones down rocky mountains and wore them down into smaller rocks and pebbles. After millions of years, a layer of very tiny pieces of rock built up on top of the earth. Pieces of dead plants and animals got mixed in with the bits of rock. This mixture is soil.

What is clay?

Clay is a special kind of soil. It is made up of fine bits—grains—of rock, tinier than those in most other kinds of soil. Many of the grains are bits of certain minerals called clay minerals. Kaolinite (KAY-uh-luh-nite) is the most common one. Clay minerals cause the clay grains to pack together tightly so they can be molded into any shape at all. When clay is baked, it gets hard. So clay is used to make bricks, pots, dishes, and other useful things.

How many oceans are there in the world?

Even though we talk about the Atlantic Ocean, the Pacific Ocean, and others, there is really just one ocean. Make a small paper boat and then try sailing it around the globe in your classroom. Start the boat at any point in the ocean. Keep it going in the water. Can you find a place where your boat must stop sailing? No. That's because all the oceans of the world are really one big ocean. Somewhere in each "ocean" there is a place where the water is joined to the water of another "ocean." The ocean has no end.

How was the ocean formed?

The earth did not always have a great ocean, as it does today. Many millions of years ago, the earth was very hot. Some scientists believe that at that time, most of the earth's water was trapped deep inside its rocks. Over millions of years, the rocks began to cool and harden. As they got hard, their water came out. It ran into the low places in the earth's crust and made the first oceans. Other scientists think the water came from great clouds that were all around the earth. As the hot earth cooled, the clouds cooled, too. Clouds that get cooler form raindrops. So the clouds around the earth rained for hundreds of years, filling the earth's low places.

Since then, the number, shape, and size of the oceans have changed. But an ocean is still a large, low area filled with water.

Where is the deepest part of the ocean?

The deepest spot is in the Pacific Ocean, near the Mariana Islands. Here the water is over 36,000 feet (10,900 meters) deep—or nearly 7 miles (11 kilometers) from the surface to the bottom. This is deep enough to swallow the highest mountain in the world—Mount Everest—which is nearly 6 miles (10 kilometers) high.

! If you were to drop a rock the size of your head into water 36,000 feet (10,900 meters) deep, it would take the rock about an hour to reach the bottom!

Why is the ocean salty?

The ocean tastes salty because there's a lot of salt in it. Most of this salt came from the land. Salt is a mineral that can be found in many rocks and in soil, too. When rain falls on the earth, some of it soaks into the ground. Some of it also trickles over rocks and along the soil. The flowing water picks up some salt and carries it to rivers. Rivers flow into the ocean. And with them goes a little salt. Because this has happened for millions of years, the ocean now has a great deal of salt in it. Rivers are far less salty, because they keep dumping their salt into the ocean—where it stays.

41

What is a salt lick?

A salt lick is a chunk of rock salt that animals like to lick. Salt is found in large rocky "beds," or layers, in the ground. These layers are made up of tiny salt grains packed tightly together so that they form rock. Long, long ago, salty ocean water lay where the salt beds are now. But the water dried up and left the salt. In a few places, part of a salt bed will stick out of the top of the earth. Such bare rocks of salt are called salt licks. Animals come there to lick salt because their bodies need it. Sometimes farmers put a chunk of rock salt in a cow pasture. This chunk is also called a salt lick.

What are tides?

Did you ever sit on a sandy beach and watch the ocean water move closer and closer to you? If you did, you were watching the tide come in. If you were still on the beach later that day, you saw the tide go out again. That means that the ocean water moved back. Once again you could see the sand that the water had covered.

In most places in the world, tides go in and out this way twice each day. They do so because the ocean water rises and falls. The rise and fall are caused by gravity—the great invisible force that all stars, planets, and moons have. The force of gravity pulls things toward the center of the star, planet, or moon. The gravity of the sun and the gravity of the moon both pull on the earth's ocean water, causing tides. The moon is much nearer the earth than the sun. So the moon's pull on the ocean water is the stronger one. Of course, the moon pulls on the earth's land as well. But the land is solid. The moon's pull isn't strong enough to move it much. Ocean water is liquid, and so it moves a lot when the moon's gravity pulls on it.

Do lakes and rivers have tides?

Yes, they do. But their tides are usually too small to be noticed.

What makes waves in the ocean?

Waves are ridges, or swells, of water on top of the ocean. They travel one after another across the sea. Most waves are caused by wind blowing over the top of the water. When wind begins to blow over a smooth stretch of water, it makes little ripples. If the wind keeps on blowing in the same direction, the ripples grow bigger. They get to be waves. The longer and harder the wind blows, the bigger the waves get.

What is a tidal wave?

A tidal wave has nothing to do with tides. It is a gigantic wave caused by an earthquake under the ocean. The quake pushes a part of the sea floor up or down and starts a long wave. The wave travels very fast, sometimes hundreds of miles an hour. As it travels it grows. At first, a tidal wave may be only a few feet high. But by the time it reaches land, the water has piled up much higher. A tidal wave can grow to be 100 feet (30 meters) high. When it hits the shore, it causes great damage. Today scientists all over the world call a tidal wave by its Japanese name—tsunami (tsoo-NAH-mee).

SURELY HE JESTS!!

Where do rivers come from?

Rivers start with rain. Wherever rain falls or snow melts, some water flows downhill. No place on the earth is perfectly level, or flat. There is always a little slope. Water moves down the slope toward the lowest place. The flowing water carves out ditches in the ground. With every new rainfall, the ditches get deeper and wider. They become streams, which flow into other streams. They grow bigger and bigger until they become rivers that flow all the time. Rivers keep flowing until they pour into the ocean.

The Nile River in Africa is more than 4,000 miles (6,400 kilometers) long. That's a lot longer than the distance across the United States from Maine to California!

What is a glacier?

A glacier (GLAY-shur) is a huge heap of ice and snow so heavy that its own weight moves it downhill. Sometimes glaciers are called "rivers of ice." Like rivers, they keep moving downhill until they reach the ocean—unless they melt first. Glaciers move very slowly. Small ones may move only an inch or so a day. Large glaciers may move as much as 10 feet (3 meters) per day.

Where do glaciers come from?

In some parts of the world, a lot of snow falls. If the temperature never gets warm enough for all of this snow to melt, new snow piles up on top of old snow. As years go by, the growing heap of snow gets thicker and heavier. Gradually most of it gets packed down into ice. When the heap gets very, very heavy, it begins to slide downhill. It becomes a moving glacier.

What is an iceberg?

An iceberg is a mountain of ice floating in the sea. It was once part of a glacier. But it broke off when the glacier reached the edge of the ocean. Although the ocean is salty, the ice of an iceberg has no salt in it. An iceberg is made of fresh water that comes from the glacier.

An iceberg is born in a very cold place. It floats out to a warmer area where it begins to melt. For about three years, it keeps traveling and melting little by little. Eventually, it gets very soft, breaks into pieces, and melts away completely.

Some icebergs are about as long as a football field. Others can be as long as 2,900 football fields lined up in a row. Most of an iceberg's ice is hidden underneath the water. What people see is just the tip of the ice sticking out above the water.

AN ICEBERG IS BORN LIKE SO.

GLACIER
LAND
WATER
GLACIER
LAND
ICEBERG
WATER

Why is an iceberg dangerous?

An iceberg is dangerous to ships at sea because it is mostly hidden under water. When a large iceberg is sighted from a ship, no one aboard can tell how far the ice extends under the water. The ship could ram itself on this hidden ice and sink. The only sure way for a ship to be safe from icebergs is to sail away from them.

An iceberg can weigh a million tons!

Weather and Climate

What is weather?

When you talk about weather, you are really talking about the air. Weather is what the air is like in any one place at any one time. How hot or cold is the air? How much dampness, or moisture, is in it? How fast is the air moving? How heavily does it press on the earth? The answers to these questions tell about the weather.

What's the difference between weather and climate?

Weather tells what the air is like in a place at any one time. Climate tells what the weather is like in general, all year round. If a place has much more dry weather than wet weather, we say it has a dry climate. If it has much more hot weather than cold weather, we say it has a hot climate. Yuma, Arizona, for example, has a hot, dry climate. On most summer, spring, and fall days in Yuma, the weather is dry, sunny, and hot. But on a winter morning, the weather may be rainy and cool. Later that same day, the weather may be dry, sunny, and cool. Weather keeps changing each day. Climate stays much the same one year after another.

49

How many different climates are there in the world?

Each place in the world has its own climate. But many climates are so much alike that scientists have grouped them all into just twelve types. Each type describes how hot or cold and how dry or wet a place is.

The United States has ten of the twelve types of climate. They range from the climate of Miami to the climate of Alaska. Miami is almost always very warm and is rainy half the year. Parts of Alaska are always cold and fairly dry.

What makes climates different?

The location of a place on the earth decides its climate. If you live far to the north, where the red arrow is pointing, you live in a cold climate. The same would be true if you lived very far to the south, where the green arrow is pointing. The sun's rays hit these areas at a great slant and don't warm the land very much. But if you live somewhere around the middle of the earth—near what we call the equator (ih-KWAY-tur)—your hometown probably has a climate that is hot all year round. That is because the sun's rays hit this area fairly directly. The more directly the sun's rays hit a place, the warmer that place is. If you live near the equator, your hometown not only gets more sun, but it also gets more rain than places very far north or south.

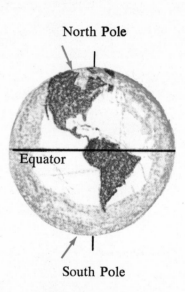

North Pole

Equator

South Pole

IF YOU LIVE AT THE EQUATOR, THE CLIMATE IS HOT.

How high up you live also makes a difference in the climate. If you live in the mountains, you are likely to have a cooler climate than if you lived lower down.

If you live near the ocean, your winters are probably less cold and your summers less hot than those in places far from the ocean. But your hometown usually has more rain than those inland places do. Winds and the movement of water in the ocean near your home help to make the climate the way it is.

Are there really "poles" at the North Pole and South Pole?

No. The picture of a pole labeled "North Pole" is just a joke. There are no poles of that sort marking either the North or the South Pole. The word "pole" here means something very different from "a long, thin stick."

The earth is round like a rubber ball. If you were to stick a long pencil through the center of a ball, one end of the pencil would stick out at each end of the ball. In much the same way, scientists have stuck an imaginary line—instead of a pencil—through the earth. The line is called the earth's axis. It is around this axis that the earth turns. The places where the ends of the axis stick out are the North Pole and the South Pole. Circling the earth halfway between the poles is an imaginary line that we call the equator.

Are the North Pole and the South Pole exactly alike?

No. The point we call the South Pole is on an ice-covered land. The point we call the North Pole is on an ice-covered sea. Water does not cool off as much as land does. So the North Pole does not get quite as cold in the winter as the South Pole. The North Pole is cold enough for most of us, though. A temperature of 73 degrees below zero has been recorded nearby!

51

A submarine has sailed right under the ice at the North Pole!

What does "below zero" mean?

When you talk about "below zero," you are talking about a temperature as measured on a thermometer. One kind of thermometer has a line of liquid that moves up and down a tube. Along the tube are numbers. The space between two numbers is called a degree. By seeing how far up or down the tube the line of liquid has moved, you can tell how hot or cold something is. Thermometers are often used to measure the temperature of the air or of water.

People in the United States most often use a thermometer that was invented by Gabriel Fahrenheit (FAIR-in-hite). On his thermometer when the line of liquid reaches 212 degrees, water boils. It freezes and turns to ice at 32 degrees. We write this temperature as 32°F. The "F." stands for Fahrenheit. On a day when the thermometer reads 32°F., you will probably wear a coat, a hat, and gloves when you go outside. At zero degrees (0°F.) you will want to bundle up in a coat, hat, scarf, gloves, and a few sweaters. Any temperature below zero is even colder, and you would probably not want to go outside at all.

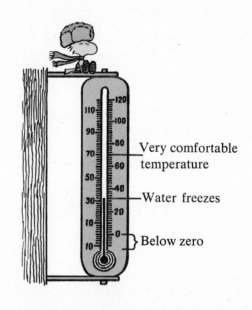

Very comfortable temperature

Water freezes

Below zero

A Fahrenheit thermometer

Are there any other thermometers besides Fahrenheit's?

Yes, there are a few more. One of these was invented by a man named Anders Celsius (SELL-see-us). On his thermometer, water freezes at zero degrees. We write this as 0°C. The "C." stands for Celsius. Water boils at 100°C. Those two numbers are easy to remember. That's probably why the Celsius thermometer is used almost all over the world.

What place has the lowest temperature on record?

The place with the coldest temperature on record is near the South Pole. At a weather station called Vostok, 400 miles (640 kilometers) from the Pole, the temperature has gotten colder than 126 degrees below zero (−126°F., or −88°C.). No people live around the South Pole, except scientists who visit the weather station. But some people make their home in another very cold place—northern Siberia, in Asia. Temperatures there have gone down as low as 94 degrees below zero on the Fahrenheit thermometer (−70° C.)!

What place has the hottest temperature on record?

The hottest temperature ever recorded was in the country of Libya in North Africa. There, in one place in 1922, the temperature reached more than 136°F. (58°C.)! If you look at the air thermometer in your house or school, you'll see that the numbers don't even go that high!

Are deserts always hot?

No! Lucy may think that the promise "I will love you till the sands of the desert grow cold" means forever. But it really doesn't. The sands of deserts *do* grow cold, almost every night. The sun warms them during the day. But at night they can't keep their heat. In most places, moisture in the air acts like a blanket, holding the heat under it and keeping the earth warm at night. But air in the desert has less moisture in it than it does in most other places. So as soon as the sun disappears in the evening, the heat from the desert floor escapes into space.

53

Why are deserts so dry?

Deserts are dry because they get very little rain. Many deserts are separated from the sea by mountains. Winds that blow onto the land from the sea carry a lot of moisture. When they start blowing up mountain slopes, these winds become cooler. Cooler winds cannot hold as much moisture as warmer ones. So the cooled-off winds drop their moisture in the form of rain or snow before reaching the mountaintops. By the time the winds reach the other side of the mountains, almost no moisture is left in them. The land on the other side of the mountains gets very little rain. It becomes a desert.

The Atacama Desert in Chile has had no rain for more than 400 years!

Which place on earth gets the most rain?

A spot on the Hawaiian island of Kauai (kah-oo-AH-ee) gets about 460 inches (1,168 centimeters) of rain each year. That's at least 400 inches (1,016 centimeters) more than most other places in the United States. If you filled a long tube with 460 inches of rain, the water would rise as high as a four-story building!

... AND NOW FOR MY SINGIN' IN THE RAIN ACT !!!

What makes the winds blow?

The air around us is always moving. It moves because the air temperature is different in different places. When air gets warmed by the sun, it gets lighter. It rises and then moves to a spot with colder air. The colder air sinks and then moves to the warm area. You feel this movement as wind.

There are two kinds of winds. One kind blows within a small area. For example, the air in a cloudy place is cooler than the air in a sunny place. The temperature difference causes the air to move, or the wind to blow.

The planetary (PLAN-ih-ter-ee) winds are the second kind of wind. They blow over large areas of the earth, and they blow all the time. Basically, they move between the cooler parts of the earth near the North Pole and South Pole and the warmer parts of the earth near the equator. Planetary winds move clouds and storms from one place in the world to another.

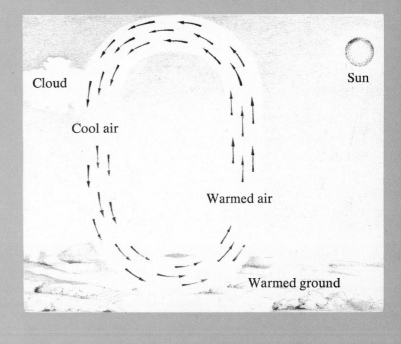

Cloud

Sun

Cool air

Warmed air

Warmed ground

How fast can winds blow?

Near the ground, winds usually blow more slowly than a car on a highway—less than 50 miles (80 kilometers) an hour. High up in the air, winds blow faster. Wind speeds of up to 231 miles (370 kilometers) an hour have been measured at the top of Mt. Washington, in New Hampshire. That is about twice as fast as a taxicab could go if it were in a race. The fastest wind ever measured was in a tornado. It was moving at 280 miles (448 kilometers) an hour.

What is a tornado?

A tornado is a noisy windstorm that often sweeps across parts of the United States. It looks like a long sleeve reaching down from a huge, dark cloud. In a tornado, wind whirls around and around in a circle about the size of two or three football fields. Very little air is in the center of this circle. Like a giant vacuum cleaner, the tornado can suck up anything in its path. Up go houses, cars, animals, people, and even railroad tracks. They may come down again later, far from where the storm picked them up. That's what happened to Dorothy and her dog, Toto, in *The Wizard of Oz*. A tornado can also flatten big buildings or even make them explode.

The whirling wind of a tornado can spin as fast as 280 miles (448 kilometers) an hour. But the whole tornado, spinning like a top, moves along at 20 to 40 miles (32 to 64 kilometers) an hour—about the speed of a car traveling down a city street.

What is a hurricane?

A hurricane is a wild windstorm that starts at sea. Like a tornado, a hurricane is made up of whirling winds. But unlike a tornado, a hurricane is very large. It usually stretches across 300 or 400 miles (480 or 640 kilometers) at one time. These two pictures show the difference in size between a tornado and a hurricane.

Inside a hurricane, the wind is whirling at speeds of from 75 to 200 miles (120 to 320 kilometers) an hour. That's not quite as fast as the winds of a tornado, but it's fast all the same. A hurricane's wild winds cause huge waves to form on the ocean. The wind and waves can sink ships. They can tear up trees and buildings on islands and on mainland seashores. Hurricanes usually bring heavy rains, too. These rains, as well as the high waves, can cause great floods. People and animals are sometimes drowned in them.

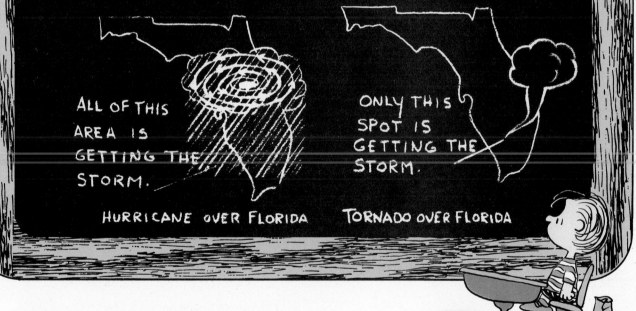

ALL OF THIS AREA IS GETTING THE STORM.

ONLY THIS SPOT IS GETTING THE STORM.

HURRICANE OVER FLORIDA   TORNADO OVER FLORIDA

What is the "eye of the storm"?

At the center of a hurricane's circle of whirling winds is a quiet space with clear skies above. This is the "eye" of the hurricane. It is usually about 20 miles (32 kilometers) across. Some people think the hurricane is over when the "eye of the storm" reaches them. The wind dies down. The sky is bright above. But the whole storm circle is still traveling. Within a couple of hours the other side of the whirlwind will arrive. It will bring more wild winds and heavy rains.

57

Where do puddles go when they dry up?

After a rain you usually see puddles in the street. A few hours later, the puddles are gone. What happened to them? They evaporated (ih-VAP-uh-rate-ed). This means that the water went up into the air and became part of it. When water is in this form we call it water vapor.

There is always some water vapor in the air. You cannot see it because it is in very tiny bits called particles (PAR-tih-kulz), which are scattered far apart from each other. The particles are so small that you could see them only under a microscope.

Water vapor comes from many places, not just from puddles. Water is always evaporating from hot pots over fires and from ponds, streams, rivers, lakes, and oceans.

What is humidity?

Humidity is the moisture, or water vapor, in the air. If there is a lot of water vapor in the air, the humidity is high. If there is very little water vapor in the air, the humidity is low.

What makes a cloud?

A cloud is made up of very tiny drops of water, called cloud droplets. Air always has some water vapor in it. If the air is warm, it is light, and it rises. As it rises it gets cooler. Cool air cannot hold as much water vapor as warm air. Particles of water vapor join together, or condense. They usually condense around tiny specks of dust or salt in the air. They form water droplets. If the air is very cold, they form bits of ice called ice crystals. The water droplets and ice crystals are light enough to float in the air. Any one droplet or ice crystal is too small for the eye to see. But a whole crowd of them make a cloud.

YES!!

WHAT IS THIS, WHAT IS THIS... IS IT SMOG OR IS IT FOG?

What is fog?

A cloud that forms close to the ground is called fog. If you walk in fog, you cannot see separate little droplets, but you can often feel them on your face. A whole crowd of droplets can make such a thick fog cloud that you cannot see through it.

What is smog?

The word "smog" is a combination of the words "smoke" and "fog." And that's pretty much what smog is.

The air always has some bits of dust floating around in it. In cities, other particles are also in the air—soot and smoke from chimneys, chemicals from factories, and fumes from automobile exhausts. We say such air is polluted. On breezy days, moving air carries the polluting particles away. On still days, a blanket of air heavy with moisture may hang over the city. Then none of the dirty particles blow away. Water droplets form around them. The cloud or fog they make is not colorless. It is dark. It is smog. When you breathe in a lot of the dirty particles all at once, they can hurt your lungs. Smog is the worst kind of air pollution.

OOPS...EXCUSE ME. I THOUGHT YOU WERE SNOOPY!

WEIRDO!!

EVEN IN A FOG, I FAIL TO SEE THE RESEMBLANCE!!

Where does dew come from?

Dew is moisture from the air that has gathered in drops on leaves and blades of grass. During the day, the sun warms the earth. But at night, the earth and the air near it usually cool off. So do grasses and other plants. Cool air cannot hold as much moisture as warm air can. So some of the moisture in the air condenses into drops of water on the leaves and grass. These drops are dew.

What is frost?

Frost is like dew. But when the night is very cold, the moisture in the air forms ice, or frost, instead of water, or dew. Like dew, frost forms on the grass and on other plants.

Does "Jack Frost" really paint ferns on your window?

No. Jack Frost is a make-believe person. The ferns that appear on the inside of your window on cold nights are really ice crystals, or frost.

At night, before the "ferns" appear on the window, the glass is warm from the warm air inside your house. The air around the glass is warm, too. But then the outside temperature drops quickly below the freezing point. It makes the glass freezing cold. The freezing-cold glass cools the air next to your window, inside your house. The cooled air cannot hold as much water vapor as the warmer air. So the water vapor forms frost on the window. If the temperature of the window were above freezing, droplets of water like dew would form on the glass. It would become "steamed up."

60

Why can you see your breath on a cold day?

If you go outside on a cold day and blow out your breath, you can see a small steamy cloud in the air. Your breath has moisture in it. Your breath is warm because the inside of your body is warm. When you blow that moist, warm breath into the cold out-of-doors air, your breath suddenly cools. Some of the moisture you have breathed out turns to water droplets. They form a small steamy cloud.

HEY, BIG SISTER... GOT NOTHING TO SAY? HAVE YOU LOST YOUR BREATH?

I'LL SHOW YOU BREATH!!! WHO MADE THAT SNOWWOMAN? WHO... WHO... WHO??!

Where does rain come from?

Rain comes from clouds. When a cloud grows big, the cloud droplets in it begin to bump into one another. They join together and form big drops. The big drops are too heavy to float in the air. They fall to earth as rain.

IT'S RAINING CATS AND DOGS?! ...SOMETIMES WOODSTOCK'S EXPRESSIONS CAN GET PRETTY GROSS!

RAIN!

IT HELPS THINGS TO GROW... IT FILLS UP THE LAKES AND OCEANS SO THE FISH CAN SWIM AROUND AND IT GIVES US ALL SOMETHING TO DRINK...

WOODSTOCK DOESN'T CARE WHAT IT IS AS LONG AS HE UNDERSTANDS IT.

61

Raindrops are not tear shaped.
They are perfectly round!

Why do we need rain?

At one time or another, we have all wanted the rain to go away and even to stay away. Rain can spoil a picnic or a ball game. But rain is very important to our lives.

Without rain, plants would die. They need water to live and grow. Animals, including people, would die, too. They would have no water to drink and no plants to eat. Nothing on earth could stay alive without rain.

AND NOW, HERE'S THE WORLD WAR I FLYING ACE LENDING HIS TALENTS TO MODERN SCIENCE.... SPRAYING CLOUDS WITH CHEMICALS TO MAKE RAIN. RAIN, YOU FOOL CLOUD.. RAIN... RAIN!!!

Can people make rain fall?

People have made rain fall by spraying clouds with a chemical that helps raindrops form more quickly. Spraying clouds in this way is called "seeding." Seeding does not always work well. In dry lands where there are no clouds to spray, it cannot work at all.

Some people pray for rain or they do rain dances. Sometimes rain follows the dance or the prayers. But scientists don't believe that such methods really work.

What causes thunderstorms?

We have thunderstorms when big, fluffy-looking clouds called thunderheads tower very high into the sky. They look beautiful when you see them at a distance. When the sun shines on their high-piled puffs, they look white. But as they sweep over your head and shut out the sunlight, they look very dark.

Thunderheads build up on hot, damp days when the very warm ground heats the moist air above it. The air rises higher and faster than usual. Water droplets gather into very big clouds. Some are several miles high! Inside each cloud, the warm rising air cools quickly. The cooled air sinks to a lower part of the thunderhead. There, the air is warmed again, and it rises. This rising and falling air makes violent winds inside the cloud. Large raindrops form. Lightning flashes, and thunder crashes.

What is lightning?

Lightning is a flash of electricity in the air. There is electricity everywhere—in clouds, in the earth, even in you! Sometimes when you walk across a carpet and touch someone, you feel a tiny spark of electricity jump between the two of you.

In towering thunderclouds, a lot of electricity builds up. As clouds draw near to one another, huge sparks or flashes of electricity pass between two clouds. Or a flash may go from a cloud to the earth. The electricity heats the air along the path of the flash so much that the air glows. That glow is what we call lightning.

Each second of every day about 100 bolts of lightning strike some part of the earth!

What is thunder?

When air is heated, the very tiny particles that make it up begin to move faster. The electrical flash from a thundercloud suddenly heats the air so much that all the particles move around wildly. The air shakes as huge numbers of them suddenly rush apart to get more space for their "dance." When this sudden huge movement in the air reaches our ears, we hear a thunderclap.

63

Why do you see the lightning before you hear the thunder?

Light travels faster than you can imagine— 186,282 miles (299,792 kilometers) in one second! So you see the glow of lightning the instant it flashes, even though it may be miles away. Sound travels much more slowly. It takes the sound of thunder nearly five seconds to travel one mile. So if a lightning flash is several miles away, you see the light right away. Then the sky darkens again. And after a pause you hear the thunder.

How can you tell how far away lightning is?

Start counting as soon as you see a lightning flash. Count "one-Mississippi, two-Mississippi, three-Mississippi," and so on. Each number will take you about a second to say. Stop counting when you hear the thunderclap. That will give you the number of seconds the sound has taken to reach you. Allow five seconds for each mile (three seconds for a kilometer), and you can tell about how far away the lightning flash was. If you counted to "five-Mississippi," the flash was about a mile (a kilometer and a half) away. If you counted to ten, the flash was about two miles (three kilometers) away.

I DON'T BELIEVE HIM!!

Can thunder and lightning hurt you?

Thunder can't hurt you, but lightning can. Thunder is just air shaking very hard. Lightning is electricity. A very small flash of electricity can give you a shock. A lightning flash is huge. It can burn whatever it touches, sometimes very badly.

Lightning usually strikes the highest thing around. This may be a skyscraper in a city, a tall tree in an open field, or a sailboat mast on the water. Metal lightning rods or specially wired television antennas can lead the electricity safely to the ground. They can keep a building safe from lightning damage. A metal car or airplane body can protect people inside it, too. However, if you stand under a big tree you will not be protected. The tree may be hit by the lightning, and the tree can fall on you. So if you are out in the open during a thunderstorm, you will be safest lying flat on the ground!

♪♪ HAIL, HAIL, THE GANG'S ♪ ALL HERE.... JUST THOUGHT I'D LIVEN UP THE DAY.

What is hail?

Hail is made up of small lumps of ice that sometimes fall to earth during thunderstorms. These icy stones are formed inside the thunderclouds.

The tops of tall thunderhead clouds are always very cold. Down near the bottom of the clouds, the air is much warmer. Inside these clouds, warm air moves swiftly up and cold air moves swiftly down. Sometimes raindrops are blown up to the freezing-cold part of the cloud before they fall. There they turn to ice. Then they are blown down again and are coated with more raindrops. Before they fall to earth, the bits of ice may be blown up and down many times. Each time more raindrops gather on them and then freeze, to form extra layers of ice on the lump. Each lump of ice is called a hailstone. If many layers gather and freeze on one hailstone before it falls to the ground, it may grow quite large. Hailstones can break windows or dent car roofs. They can flatten plants in fields and gardens.

> **!** Hailstones as big as your head have fallen.
> Some have measured 17 inches (43 centimeters) around.
> One weighed 1½ pounds (680 grams)! **!**

What is sleet?

Sleet is frozen rain. It falls when the air close to the ground is freezing cold. Sleet starts out as rain. As the raindrops fall, they freeze. They form the tiny ice balls known as sleet.

GOOD GRIEF, THEY'RE PREDICTING RAIN TURNING TO SLEET!

What is an ice storm?

On some cold days, rain falls and turns to ice after it has landed. The frozen rain forms a smooth, slippery coat of ice on the freezing-cold street. We call this kind of rain freezing rain or an ice storm.

THINK OF IT THIS WAY... ONLY THREE MORE MONTHS TILL SPRING TRAINING.

Is snow frozen rain?

No. Raindrops that freeze as they fall form sleet, not snow. Snowflakes are formed right in the clouds.

Clouds floating in freezing-cold air are made of tiny crystals of ice. As the air grows colder, more and more water vapor condenses around the ice. The tiny crystals grow bigger and bigger. The snowflakes you see are simply these crystals after they have grown too large and heavy to float in the air. They fall to earth as snow.

What do snowflakes look like?

If you look at a group of snow-flakes under a magnifying glass, you will see that they are like beautiful small lace doilies. They have many different sizes, shapes, and lovely patterns. However, if you count their sides, you will find that each of them has six. Each snowflake has six points, too.

No two snowflakes are exactly alike!

Can you count all the snowflakes that fall?

Linus and Lucy say they have counted the snowflakes. This is a joke. No one could count all the snowflakes that fall. There are too many, and no one can see them all. Not Linus, not Lucy—not even you!

69

How do the weather forecasters know what tomorrow's weather will be?

Tomorrow's weather is already building up in the air above the earth. Weather forecasters get reports on what is happening in the air all around the world. Many thousands of weather stations all around the world send out messages to weather forecasters. These stations measure the amount of rain or snow that falls in their area. They keep track of heat or cold with thermometers. They measure how heavy the air is, and how much moisture it holds. They find out how fast the planetary winds are carrying the weather and in what direction. Airlines and ships at sea send radio messages every few hours about the weather where they are. Cameras and other equipment circle the earth in weather satellites. They send back pictures and other information. All these facts are put together on special maps. The maps show what kind of weather is heading your way, wherever you are.

70

What are cold fronts and warm fronts?

A cold front is the leading edge of a moving clump of cold air called a cold air mass. An air mass is a large blob of air that stays together as it moves across the earth. The entire blob has about the same amount of moisture and temperature. Its temperature and moisture depend on where it comes from. A cold front often brings showers and thunderstorms with fast winds.

A warm front is the leading edge of a warm air mass. It often brings along steady rains or snow.

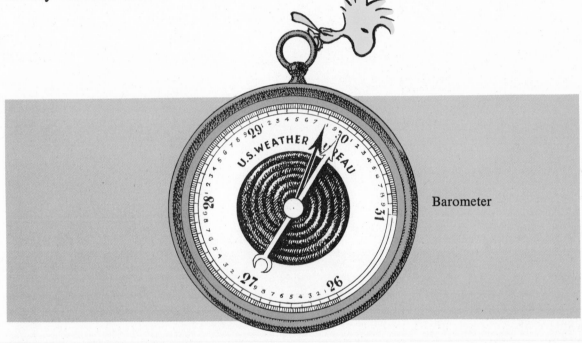

Barometer

What do weather forecasters mean when they say the barometer is rising?

In a weather report the weather forecaster may say that the barometer (buh-ROM-uh-tur) is rising or falling. A barometer is a special instrument that measures how heavily the air is pressing on the earth. When the barometer is rising, it means that the air is pressing harder and harder on the earth. When the barometer is falling, it means that the air is pressing less and less on the earth.

Knowing the air pressure helps people predict the weather. When the air pressure is rising, clear skies and cool weather are probably on their way. When the air pressure is falling, stormy weather is probably in store for us.

The word "High," circled on a weather map, shows the center of high air pressure. The word "Low," circled on a weather map, shows the center of low air pressure.

What good are weather forecasts?

Weather forecasts tell what weather is coming. They help people know what kind of clothes to wear. They warn farmers against frosts that might ruin their crops. Weather forecasts warn people along seacoasts against hurricanes, and in other places against tornadoes, so they can close up their homes and get to safe shelters. In the winter, forecasts tell skiers how much snow will fall. During the summer they tell people how warm and sunny the beaches will be. They warn when smog will be bad in cities. They say when roads will be slippery or foggy and dangerous for driving. They help airplane pilots choose safe routes to fly.

Unfortunately, the weather doesn't always work out the way it's been forecast. Winds sometimes change direction. They blow storm clouds to unexpected places. Sometimes winds speed up or slow down. Then weather changes reach an area later or sooner than forecast. Nevertheless, the weather forecasters do a lot to help people keep comfortable and safe.

The Stars and the Planets

THEY SAY THERE ARE AS MANY STARS IN THE SKY AS THERE ARE GRAINS OF SAND ON ALL THE BEACHES IN THE WORLD.

LOOK...THE NORTH STAR.. MAKE A WISH.

HOW ROMANTIC!!

SIGH

What are stars?

When you look up at the sky on a clear night, you see many, many twinkling points of light. These are the stars. They are really huge balls of bright, hot, glowing gases. They pour out light, just the way the sun does. In fact, the sun *is* a star. The other stars look much smaller than the sun because they are much farther away from the earth.

What makes the stars shine?

Stars shine with their own light because they are very hot. A lot of gases press down at the center of a star to cause this great heat.

Why do stars twinkle?

Stars don't really twinkle. They just seem to. There is a thick blanket of air around the earth. The light coming from the stars must pass through this air. As the starlight passes through, it shifts, or moves about. This happens because of moisture in the air, changing air temperatures, and the constant movement of the air. To us the shifting starlight looks like twinkling.

Long, long ago, some people believed the stars were lamps hanging from a huge ceiling!

How many stars are there in the sky?

On a clear night, you can see about 2,000 stars just by looking up. You could see many thousands more with a small telescope. You could see billions more with a very powerful telescope. Scientists keep inventing stronger telescopes. With each new one, the scientists discover more stars. So no one really knows how many stars there are in the sky.

What does a telescope do?

A telescope makes things appear much larger than they really are. In this way it is something like a magnifying glass, but much more powerful. A telescope also makes things appear brighter than they are. When you look through a telescope, faraway things seem closer and look clearer. So for a long time people have been using telescopes for studying the stars and planets.

Today there is another kind of telescope that has nothing to do with seeing. It is called a radio telescope. It picks up movements in the air called radio waves that come from outer space. All stars and some planets give off these waves. So do other faraway objects. By listening to radio waves, scientists learn more about everything in outer space.

Will the same stars always be in the sky?

No. Old stars are always dying and new stars are always being born. Some stars last only a few million years. Others go on and on for hundreds of billions of years. But all stars either explode or get small and stop shining. At the same time, new stars keep forming from gas and dust in space.

> **!** It takes many millions of years for light to travel from a far-off star to your eyes. So some of the stars you see in the sky really burned out long, long ago! **!**

Where do the stars go in the daytime?

They don't go anywhere. They are always in the sky. But the sun's bright light keeps you from seeing the stars during the day. Only in the evening can you begin to see them again.

How big are stars?

Stars are the biggest single thing scientists know of. Our own sun is a star, and it is more than a million times larger than the earth. The sun is a middle-sized star, and many other stars are about its size. But some stars, called dwarfs, are much smaller—about the size of the earth. Other stars, called giants and supergiants, are much larger than the sun. The biggest supergiants known are about 500 times larger than the sun.

What is the Big Dipper?

The Big Dipper is a group of bright stars in the sky. They look like the cup and handle of a water dipper. The Big Dipper is one of many groups of stars that people have named for their shapes. We call these groups constellations (kon-stuh-LAY-shunz).

Is there such a thing as a double star?

Yes, there are thousands of them. A double star is made up of two neighboring stars that travel around each other. Mizar (MY-zahr), the star at the bend of the Big Dipper's handle, is really a double star. Mizar is so far away that without a telescope, the second star is very hard to see. Most other double stars look like single stars unless you look at them through a telescope.

76

How can the stars help you if you are lost at night?

One very important star can help you—the North Star. There are no South, West, or East stars—in spite of what Lucy says. You can find the North Star easily. The two stars in the front of the Big Dipper point to it.

If you face the North Star, you are facing north. On your right is east. On your left is west. Behind you is south. If your home is south, you turn around and walk away from the North Star.

Which star is nearest to earth?

The sun is the star nearest to earth. It is 93 million miles (150 million kilometers) away. That may not sound very near, but it is close enough to give the earth light and heat.

How hot is the sun?

The outer part of the sun is about 10,000 degrees Fahrenheit (10,000°F., or 5,500°C.). Any metal known on earth would melt at such a high temperature. Most other things on earth would burn up. The inside of the sun is even hotter than the outside. It is almost 30 million degrees Fahrenheit (30,000,000°F., or more than 16,000,000°C.).

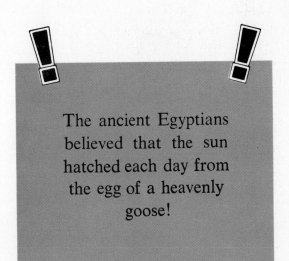

The ancient Egyptians believed that the sun hatched each day from the egg of a heavenly goose!

What is the universe?

The word "universe" means everything there is—the sun, the moon, the stars, the earth and all things on it, the other planets, and anything else you can think of. All of space and everything in space is part of the universe. It extends much, much farther than you could see with the most powerful of all telescopes. Most scientists believe that there is an end to the universe. But no one knows where that end is.

What is a galaxy?

A galaxy is a huge group of stars that are close together—close for stars, that is. They are actually many millions of miles apart! Through a telescope, galaxies look like islands. Each one has billions of stars in it. Scientists don't know how many galaxies there are in the universe, but they believe there are millions and millions of them.

All the galaxies in the universe are rushing away from each other at terrific speeds. They will keep on doing so, maybe forever!

What is the Milky Way?

If you look at the night sky, you can often see a glowing band of light. It is what we call the Milky Way. This band is made up of billions of stars. You cannot see the separate stars in the band because they are so far away from the earth.

The glowing band is actually only part of the galaxy that we call the Milky Way Galaxy. This galaxy includes all the separate stars we see in the night sky. These stars are closer to the earth than the band is, so we don't see them all blurred together. Our own star—the sun—and the earth are both part of the Milky Way Galaxy. From far, far out in space, beyond our galaxy, the whole galaxy would appear as one big band of light with a bulge in the middle. From another part of space beyond our galaxy, the Milky Way would appear as a glowing spiral-shaped island.

What is a solar system?

A solar system is a family of ball-shaped objects in space. The family consists of a star, or sun, in the middle with any number of planets traveling around it. The earth is part of a solar system that has a sun in the middle and nine large planets moving around it. The earth is one of these nine planets.

How did our solar system begin?

Scientists don't know for sure. But many think that it formed from a huge pancake-shaped collection of dust in space. For some unknown reason, the "pancake" started to spin. As it spun faster and faster, the center became very hot. It became our sun. At the same time, great blobs of dust broke off from the edges. These collected into nine ball-shaped planets.

The Milky Way from the side.

The Milky Way from above.

What are the planets in our solar system?

There are nine planets in our solar system. The closest to the sun is Mercury. Then come Venus, Earth, Mars, Jupiter, Saturn, Uranus, Neptune, and Pluto. Besides these planets, our solar system also has asteroids (ASS-tuh-roidz) in it.

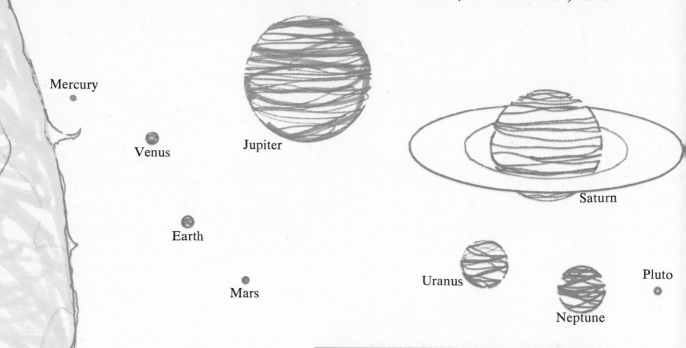

Mercury

Venus

Jupiter

Saturn

Earth

Mars

Uranus

Neptune

Pluto

What are asteroids?

Asteroids are tiny planets in our solar system. There are thousands and thousands of them. Most of them travel around the sun between the paths of Mars and Jupiter. The largest is less than 500 miles (800 kilometers) wide. Most asteroids are chunks of rock that are less than one mile wide.

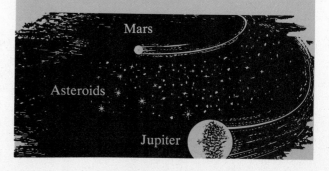

Mars

Asteroids

Jupiter

Can we see any of the nine planets without a telescope?

Yes, we can see six of them—Mercury, Venus, Mars, Jupiter, Saturn, and Uranus. Uranus is very hard to see because it is so far away. Mercury is even harder to see because it is so small. You can usually tell when you're looking at a planet in the night sky because it shines with a bright, steady light. The planets —except for Mercury—usually don't twinkle the way stars do. Whether or not a planet seems to twinkle depends on how far away it is from the earth and where it is in our sky.

The evening star we first see in the night sky is not a star at all.
It is a planet—either Mercury, Venus, Mars, Jupiter, or Saturn!

Are the other planets like the earth?

Not really. Only the earth has air to breathe and water to drink. Mercury is a dead, airless world. During the day it is extremely hot because it is so close to the sun. Venus is covered with thick clouds all the time and is also very hot. Mars is a cold world with hardly any air. Jupiter probably has no solid ground. Scientists believe it may be hot liquid all the way through. The other planets are made mostly of gases and are very cold.

Which of the planets have moons?

Seven planets have moons. Earth has one moon. So does Pluto. Mars has two, and Neptune has two. Uranus has five moons. Saturn has ten. Jupiter has thirteen. Each moon travels in a path around its planet.

How do the planets move?

Each planet moves in two ways at the same time. It moves in a huge egg-shaped path around the sun. This path is called an orbit. Sometimes the planets are closer to one another than at other times. Each planet also spins around on its own axis. A planet's axis is an imaginary line going through the center of the planet. The planet turns around this line. If you were to stick a knitting needle through the center of a ball and spin the ball, it would be spinning on its axis.

Planets orbiting the sun

THAT JUST DOESN'T WORK..

I HAVE TO SLEEP IN THE SAME DIRECTION THAT THE WORLD TURNS

Why don't the planets bump into each other?

The planets can't bump into each other because they travel around the sun in orbits that are millions of miles apart from each other. No planet ever moves out of its orbit.

The planets don't run around bumping into each other because they are polite — unlike some people I know.

83

Does the solar system move?

Yes. The sun and all its planets are traveling around the center of the galaxy. The whole solar system is moving at the speed of 175 miles (280 kilometers) a second!

THAT'S A FACT.

The huge planet Saturn is so light that it could float on water!

What are Saturn's rings?

The rings are made up of millions of tiny bits of ice and bits of ice-covered dust that orbit Saturn.

Which planet is nearest to the earth?

Venus is the planet nearest to the earth. Much of the time it is about as far from us as we are from the sun.

...AND THAT'S A FACT!

Which planet is farthest from the earth?

Pluto is the planet farthest from the earth. It is about 40 times farther from the sun than the earth is.

The sun is so far away from Pluto that Pluto never has sunlight—just the darkness of night!

Which is the largest planet?

Jupiter is the largest planet. All the other planets put together could easily fit inside it.

Which is the smallest planet?

For many years scientists believed that Mercury was the smallest planet. It is only a little larger than our moon. But now scientists have discovered that faraway Pluto is smaller—even smaller than our moon.

How big is the earth?

It is about 10 times bigger than the planet Mercury. This means that 10 planets the size of Mercury could fit inside the earth. On the other hand, the earth is about 1,300 times smaller than the largest planet, Jupiter. This means that 1,300 planets the size of the earth could fit inside Jupiter.

If you could dig a tunnel through the center of the earth from one side to the other, you would find it measured almost 8,000 miles (12,800 kilometers). That distance is more than 100,000 football fields placed end to end. If you walked clear around the earth, you'd have to walk almost 25,000 miles (40,000 kilometers)—more than 350,000 football fields.

What makes the sun rise?

As Linus said, the sun doesn't rise, the earth turns. The earth is always turning on its axis. If you have a globe at home or in your classroom, you can do an experiment to see why the sun seems to rise. Place a lamp or flashlight so that it shines on the globe. Pretend that the light is sunlight. You can see that the light is hitting only one part of the globe. Now turn the globe slowly. As the globe turns, the part that is lit changes. In the same way, as the earth turns, the part that gets sunlight changes. When the side of the earth you live on doesn't face the sun, you have night. When the earth turns farther around, the part you live on comes into the sunlight. Then the sun seems to rise in the sky. And you have daylight.

Why do we have seasons?

The earth travels around the sun. It takes a year to make the trip. The earth's axis doesn't point straight up and down. So the earth tilts, or tips, a little to one side as it travels. This tilt gives us our four seasons. When the part of the earth you live on tilts towards the sun, you get the most hours of sunlight and the most heat. Then it is summer. When your part of the earth begins to tilt away from the sun, you get less sunlight and less heat. It is fall. When your part of the earth tilts still farther away from the sun, you get even less sunlight and heat. Then it is winter. When your part of the earth begins to tilt closer to the sun again, you get more hours of sunlight and more heat again. Then it is spring.

How fast does the earth travel around the sun?

Scientists have figured out that the earth is racing through space at 66,600 miles (107,200 kilometers) an hour—thousands of times faster than the fastest racing car. During the time it took you to read this answer, the earth probably moved through space more than 300 miles (480 kilometers)!

Why don't you feel the earth moving?

You can't feel the earth moving through space because it moves so smoothly. When you ride in a car, you know you're moving, even if you close your eyes. That is because the ride is bumpy. When you are on a jet plane and you close your eyes, most of the time you cannot tell that the plane is moving. That is because the ride is quite smooth. The movement of the earth through space is even smoother. And so you cannot feel it at all.

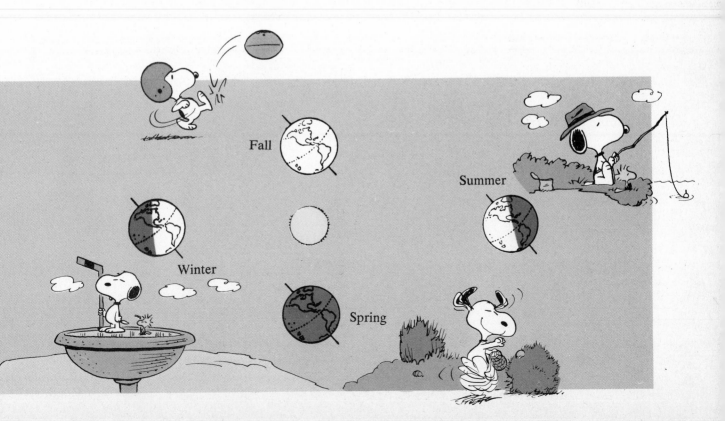

Fall

Summer

Winter

Spring

Why does the moon shine?

The moon does not shine with its own light. It has no light to give out. The moon reflects, or sends back, some of the light rays that come to it from the sun. Those light rays reach the earth—and your eyes. On a sunny day, a friend's eyeglasses reflect light in much the way the moon does. The glare you see on them isn't their own light. Instead they are reflecting the sun's rays to your eyes.

Just as the sun shines on some part of the earth at all times, the sun shines on some part of the moon at all times. The moon is always reflecting some sunlight, but you cannot always see it. During the day, the sun shines on the part of the earth where you live. The sun's light is brighter than the moon's light. So the sunlight usually hides the moon from your sight. At night no sunlight hides the moon, so you can see it "shining."

The sun's rays bounce not only off the moon, but off the earth. If you were out in space, you would see the earth shining more brightly than the moon!

How far from the earth is the moon?

The moon is about 239,000 miles (384,000 kilometers) away from the earth. How far is that?—farther than a rope would extend if it were long enough to be wrapped around the earth nine times.

Why can't we see the back of the moon?

Just as the earth moves in two different ways, the moon also moves in two different ways. It turns on its axis and it travels in its orbit around the earth. The moon takes 27 days, 7 hours, and 43 minutes to turn around once on its axis. The moon takes just about the same amount of time to travel once around the earth. This means that the moon always keeps the same side facing the earth. From the earth you never see the back of the moon.

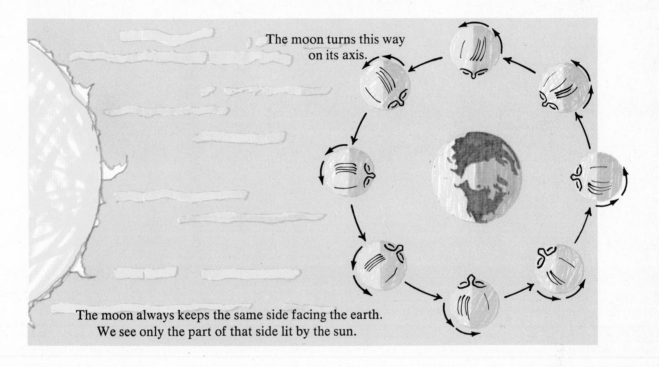

The moon turns this way on its axis.

The moon always keeps the same side facing the earth. We see only the part of that side lit by the sun.

Why doesn't the moon always look round?

The moon has no light of its own. Light comes to it from the sun, just as light comes to the earth from the sun. A part of the moon is always turned to the sun and a part of it is always turned away from the sun. One part is as dark as night. The other part is as bright as day. We see the moon only when some of the lighted part is turned toward the earth.

As the moon travels around the earth, it always keeps the same side facing the earth. But we can't always see all of that side because different amounts of it are lit by the sun during different days of the month. So sometimes we see just a sliver. Sometimes we see a half moon, and sometimes we see a full moon. The picture on this page should help you understand how this happens.

KEEP MOVING, SCOUTS, AND WE'LL GET HOME BY THE LIGHT OF THE MOON...

89

What causes an eclipse of the moon?

When light shines on anything, that thing casts a shadow. The earth is no exception. When the sun shines on the earth, the earth has a shadow on its opposite side. An eclipse of the moon happens when the moon moves behind the earth and into the earth's shadow. Because the moon is in this shadow, most of the sun's light cannot hit the moon. There is scarcely any light to bounce off the moon, so you can hardly see it. What you do see of the moon looks reddish. When the moon comes out from the earth's shadow, it shines again with full light from the sun. The eclipse is over.

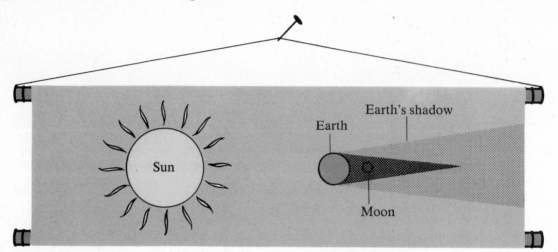

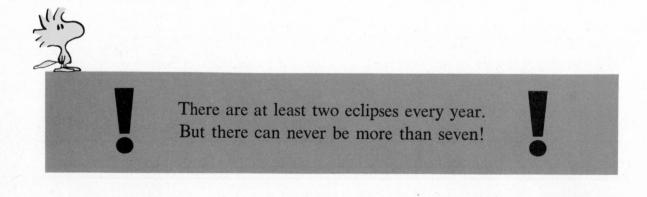

There are at least two eclipses every year.
But there can never be more than seven!

What causes an eclipse of the sun?

An eclipse of the sun happens when the moon moves directly in front of the sun. The moon's shadow is cast on the earth. The sun's light is blotted out at certain places on the earth by the moon. If you are in one of these shadowy places, you see an eclipse of the sun. You see the round disk of the moon passing across the face of the sun.

Because looking straight at the sun can damage your eyes, you should never do it—even during an eclipse.

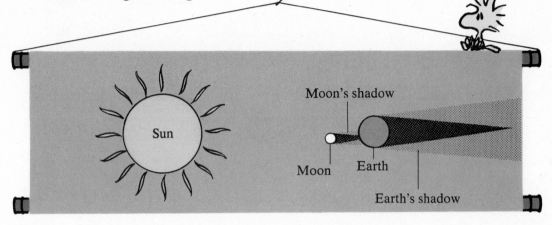

What's the difference between a "total eclipse" and a "partial eclipse"?

"Total eclipse" means that the whole sun or the whole moon is blocked from view. "Partial eclipse" means that only a part of the moon or the sun is blocked out.

Total eclipse of the sun.

Partial eclipse of the sun.

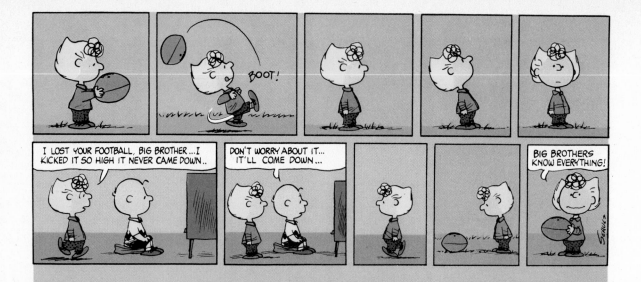

Why can't a football fall off the earth?

A football, or any other thing, cannot fall off the earth because it is always pulled to the earth by gravity. Gravity is a force—a tremendous pull—that draws all things on the earth down toward the earth's center.

Is the earth the only planet with gravity?

No. Each planet has gravity. That means that Mars pulls things toward its center. Pluto pulls things toward its center. So do Saturn, Jupiter, and all the others. In fact, everything in the universe has gravity—even a pencil and a grain of sand. Of course, the bigger the object is, the stronger its pull. A tiny asteroid doesn't have much gravity. Stars have the greatest gravity because they are bigger than any other objects in the universe. A spaceship at equal distance from a star and an asteroid would be pulled toward the star. The strong pull of our sun keeps the planets in their orbits around it.

What is a comet?

A comet is a large ball of glowing gases, dust, and ice. It travels in a long cigar-shaped orbit around the sun. Comets that can be seen without a telescope always have tails of glowing gases streaming out behind them. A comet's tail always points away from the sun because the strong energy coming from the sun blows the glowing gases backward off the comet.

TAIL

HEAD

nucleus   coma

!! A comet can be over a million miles (1,600,000 kilometers) wide!

Why are some people afraid of comets?

Snoopy and Woodstock may be afraid of comets, but most people today aren't. Long, long ago, however, comets were very frightening. People believed that a comet's sudden appearance in the sky meant that bad things were going to happen—sickness, war, poor crops, floods, the death of a ruler, or even the end of the world. They were afraid that the comet would crash into the earth and destroy it. In 1910, the earth did pass through the tail of a comet called Halley's Comet. Nothing seemed to happen because of it. However, a head-on collision between the earth and a comet could be very dangerous. Happily, no comet except Halley's has come very close to the earth. So we haven't had any problem yet, and we probably never will.

93

What is a shooting star?

Shooting stars aren't stars at all. They are the bright streaks of light you can see passing quickly through the dark sky on a clear night. Scientists call these streaks meteors (MEE-tee-urz). A meteor can be caused by a bit of dust or a very tiny rock that has been traveling at a terrific speed through space. When the bit of rock or dust hits the earth's blanket of air, its movement heats the air. The air gets so hot that it glows. The long trail of glowing air is the shooting "star" we see in the night sky. A very brilliant meteor that leaves a shining trail as it streaks across the sky is called a fireball. Its trail lasts for a few minutes.

What's the difference between a meteor and a meteorite?

A meteor is a streak, or flash, of light. You see it when a bit of rock or dust from outer space enters the earth's air at high speed. Usually the rock or dust burns up before it hits the ground. A meteorite (MEE-tee-ur-ite) is any piece of rock from space that doesn't burn up. It falls through the earth's blanket of air and lands on the ground.

What is a meteor shower?

A meteor shower happens when many, many meteors fall from the same place in the sky. A meteor shower can last for hours or even a few days. Scientists think the meteors of a meteor shower are caused by millions of tiny pieces of a broken-up comet. These pieces crash into the earth's air and burn up in it.

During a 20-minute meteor shower in 1966, people counted 2,300 meteors each minute!

QUICK, QUICK... CATCH A FALLING STAR CHARLIE BROWN...

What is astrology?

People who believe in astrology think that the stars and planets have a lot to do with their lives. Since everything in the universe keeps moving, the planets and stars are in different places on different days of the year. People who believe in astrology think that the position of certain planets and stars influence their lives in special ways, depending on when and where they were born. They say that the positions of the planets and stars tell them what kind of people they are, how they can live better lives, and what will happen in the future. The advice an astrology expert gives about the future is called a horoscope (HAWR-uh-scope). Most scientists do not believe in astrology and horoscopes.

What is meant by the "space age"?

The space age began when rocket ships were first sent out into space. In 1957, Russia sent Sputnik I into space to circle the earth. After that, a dog named Laika and a monkey named Sam became the first living space travelers.

The first person to fly around the earth in outer space was a Russian named Yuri Gagarin (gah-GAH-rin), in 1961. A month later, Alan B. Shepard became the first American to go into outer space. Since then, many other people have traveled in space. So we say that we are living in the space age.

What is space?

To most people the word "space" means the huge emptiness that is all around the earth. But space is not only out among the faraway stars. It is also right close to home. Whenever you go from one place to another, you are moving through space.

Is there an end to space?

Scientists don't know for sure. The part of space that begins about 100 miles (160 kilometers) above the earth is usually called outer space. The part of outer space among the planets is called interplanetary (in-tur-PLAN-ih-ter-ee) space. It spreads out for about 4,000,000,000 (4 billion) miles (more than 6 billion kilometers). It includes the nine planets that travel around our sun. Even farther out is deep space. That is where the stars are.

How far out does space go? There's no way to tell.

Who owns outer space?

We all do! Most of the powerful countries of the world have agreed that outer space should belong to everybody.

Why do people explore space?

People have always been curious about unknown places. At one time, people living in one place on earth knew very little about the rest of the world. But they wanted to learn as much as they could. So sailors traveled the seas and discovered new lands. Explorers journeyed through the western part of the United States and to the North and South poles. Now we can say that almost every part of our world is known.

Today there are new worlds to discover. People want to know what is out in space. They want to find out more about the planets and the stars.

Spacecraft have already landed on Venus and Mars. Even though no astronauts were on board, these spacecraft sent back scientific information and pictures. Astronauts have walked on the moon. In the future, people will be able to travel to the planets, hoping to find out still more about unknown places.

Can an airplane fly in outer space?

No. An airplane is held up by air that streams around the wings as the plane moves forward. Above the ground, the air starts to thin out. At 20 or 30 miles (32 or 48 kilometers) up, the air becomes too thin to hold up an airplane. Beyond 100 miles (160 kilometers), there is almost no air at all. So an airplane cannot fly there.

What does the sky look like in outer space?

If you were in outer space, you would see the sun, the moon, and the stars bright in a black sky—all the time. On earth, the daytime sky does not look black. The air scatters sunlight all around, which brightens up the sky and makes it seem blue. The sky around other planets may appear to be other colors. Different gases hang over each planet and scatter the sunlight in different ways. But far from a planet, there are no gases, so there can be no scattering of light. The sky looks black.

Pictures of the planet Mars show that its sky is orange-pink!

Is outer space hot or cold?

Everything whirling in space is hot or cold or in between. Stars are like huge furnaces. When heat streams out from them and reaches anything in space, that thing gets hot, too. That's why planets close to a star are hot or warm. The farther a planet is from a star, the colder that planet is. Most of outer space is not heated by stars. It is very cold—nearly 460 degrees below zero Fahrenheit ($-460°$ F., or $-275°$ C.).

Are there sounds in outer space?

No. When something vibrates—shakes back and forth quickly—in the air, movements called sound waves are sent out. The sound waves move through the air to your ears. Then you hear the sound. But outer space has no air to carry sound waves. So you cannot hear any sounds.

Huge explosions are always taking place on the sun.
If there were air all the way from the sun to the earth,
we would hear the roar of these explosions all the time!

Are there clouds in space?

There are clouds, but not like the ones near the earth. Warm, moist air floats upward from the earth and cools off. Some of that moisture then gathers into small water drops or bits of ice. Many of these drops of water and bits of ice together form a cloud. Most clouds are only a few miles above the ground.

In outer space there is no water, so clouds of moisture do not form. But there are huge clouds of gas and dust in deep space. They hide some of the distant stars from us.

The earth with clouds, as seen from space

What is radiation?

Scientists use the word radiation (ray-dee-AY-shun) for anything that flows outward—much like the spray of water from a garden hose. All radiation travels in the form of waves. An example of this is light, streaming from the sun or from a lamp. Another example is the heat that comes out of the sun. Radio and television broadcasts move through wires and through the air as streams of radiation.

Is there radiation in outer space?

Yes, outer space is crisscrossed by many kinds of radiation. There are light waves. There are x-rays—just like the ones the doctor uses to take pictures of your insides. And there are other kinds of radiation, too—all moving through space at about 11 million miles (almost 18 million kilometers) a minute.

Is radiation dangerous?

In outer space, most kinds of radiation are very dangerous. People cannot live if the radiation hits them directly. Astronauts must be protected by their spacecraft or spacesuit at all times. Here on earth, the air protects us from nearly all the harmful rays. Even so, some of them can get through and cause bad sunburn.

What are radiation belts?

All around the earth there are two invisible clouds made up of very tiny specks called electrons and protons. These specks are so small that you cannot see them, even under a microscope! The clouds they make up are called radiation belts.

Inside the belts, the radiation is deadly. It is thousands of times stronger than a person could stand. Astronauts are protected by their spacecraft if they must pass through the belts on their way to outer space. Usually they can steer their spacecraft away from the belts through escape zones.

There are very large, strong radiation belts around the planet Jupiter. Other planets may have them, too.

105

What other dangers are there in space?

There are bits of rock, called meteoroids (MEE-tee-uh-roidz), flying around everywhere in outer space. Many of them are no bigger than a grain of sand. Some move hundreds of times as fast as a rifle bullet. They go so fast that even the smallest ones can do great damage to anything they hit.

What is a spaceport?

A port is a place where ships stay before a trip. Ships also load at a port and leave from a port. So a "spaceport" is a place where "spaceships" stay before a trip, where they load, and from which they take off. A spaceport has hangars —buildings where spacecraft are kept. It also has storage tanks for rocket fuel and room for all the other equipment needed for space travel. The main American spaceport is at Cape Canaveral (kuh-NAV-er-ull), Florida.

Kennedy Space Center, Cape Canaveral

What is the difference between a spacecraft and a spaceship?

The two words mean the same thing—any rocket-powered machine that can carry people or material in space. The word "spacecraft" is usually used in talking about a real rocket-powered machine. For example, astronauts went to the moon in an Apollo spacecraft. The word "spaceship" is used mainly in science-fiction stories.

What does a spacecraft look like?

These are pictures of a few different kinds of spacecraft with the rockets that sent them into space. You can see how large they were by comparing them with the size of the man. The Apollo 5 with its rocket was 363 feet (111 meters) tall—as high as a 45-story building. It weighed more than 3,000 tons (2,700 metric tons).

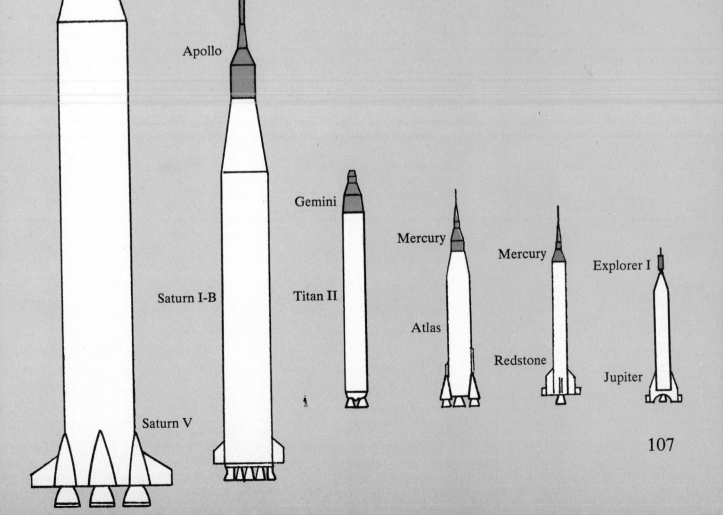

Apollo

Saturn V

Apollo

Saturn I-B

Gemini

Titan II

Mercury

Atlas

Mercury

Redstone

Explorer I

Jupiter

What is a rocket?

A rocket is a kind of engine, or motor, that is powerful enough to lift a very heavy spacecraft off the earth. In order to do this, it burns special fuels, just as an automobile burns gasoline as fuel. But a rocket may need up to ten million times more fuel than an automobile does!

Sometimes the word "rocket" is used to mean any rocket engine. Sometimes "rocket" is used to mean the main rocket engines of a spacecraft. And at other times people use the word to mean the rocket engines along with the rest of the spacecraft.

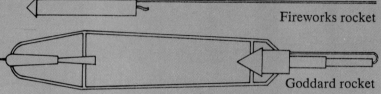

Fireworks rocket

Goddard rocket

Who made the first rocket?

Nobody knows exactly. The Chinese were using rockets more than 800 years ago. These rockets were powered by gunpowder. They were like the skyrockets that you see in Fourth of July fireworks shows.

In 1903, a Russian schoolteacher named K. E. Tsiolkovsky (tzawl-KAWF-skee) had the idea of using rockets for flights into space. In 1926, the American scientist Robert H. Goddard sent up a rocket that went about as high as a 20-story building.

How does a rocket work?

Strangely, in order for something to move in one direction, it must give a push in the opposite direction. When you row a boat, you push the water in the opposite direction from the way you want to go. When you swim in a pool, you sometimes push back against the pool wall to move you forward quickly.

This kind of two-way action is what makes a rocket motor work. Fuel is burned inside the rocket. This is called "firing" the rocket. The burning fuel forms great clouds of hot gas. The heat makes the gases swell up. They need more room, and they can escape only through an opening at the back of the rocket. So, as the gases rush out at the back, the spacecraft is pushed forward.

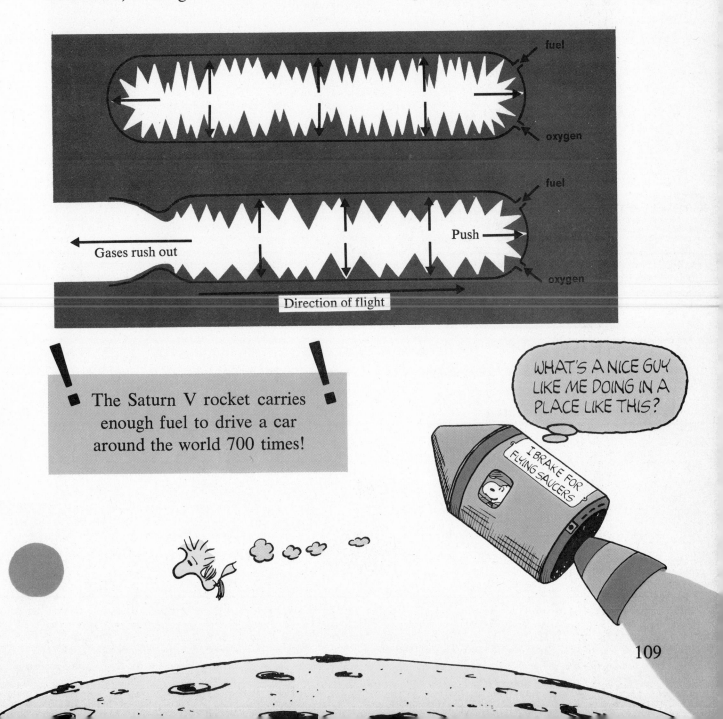

fuel

oxygen

fuel

Push

Gases rush out

oxygen

Direction of flight

! **!**
The Saturn V rocket carries enough fuel to drive a car around the world 700 times!

WHAT'S A NICE GUY LIKE ME DOING IN A PLACE LIKE THIS?

I BRAKE FOR FLYING SAUCERS

What is a countdown?

A countdown is a check-up time before a spacecraft is sent up from the earth. During this time, every inch of the rocket and spaceship is tested to see that it is in perfect working order. All the machinery that sets off and guides the ship is tested, too. A green light is switched on for each part that is in good working order. If something is not working properly, the countdown stops until that part is fixed. A person speaking over a loudspeaker keeps telling everyone at the spaceport how many hours and minutes of countdown are left.

A countdown may take hours or even days. It will continue until all the dials flash a green light. Finally the loudspeaker booms, "10—9—8—7—6—5—4—3—2—1—ZERO!" This instant is called T-zero (Time-zero). With a loud roar the rocket blasts off and the ship begins to rise.

What is an orbit?

The path of one object in space around another is called an orbit. The planets move in orbits around the sun. Each planet's orbit is shaped something like an egg.

The earth takes a year to go once around its orbit. Planets closer to the sun take less time than this. Planets farther away take longer.

Now, in the space age, we can send spacecraft into orbit around the earth, the moon, and the planets.

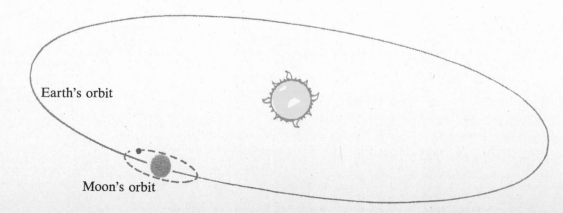

Earth's orbit

Moon's orbit

How is a spacecraft sent into orbit?

This is done in three steps. Each step is called a stage. Here's how an Apollo spacecraft would be sent into orbit around the earth before going on to the moon.

First-stage rockets are sometimes called boosters. They give the spacecraft a powerful push that lifts it from the ground. In two and a half minutes, the spacecraft is 40 miles (64 kilometers) up. It is then going 6,000 miles (9,600 kilometers) an hour. At that time the first-stage rockets stop firing. They are dropped off into the ocean. This makes the spacecraft lighter and saves power on the rest of the trip.

Next, the second-stage rockets fire for six minutes. Then they are dropped off, too. The spacecraft is now about 110 miles (176 kilometers) up, going over 14,000 miles (22,400 kilometers) an hour.

The third-stage rockets are then fired for about two minutes. This gets the spacecraft to a height of 120 miles (192 kilometers) and a speed of about 17,500 miles (28,000 kilometers) an hour. It is now in orbit around the earth, going about 30 times as fast as a jet plane.

The third-stage rockets stay in place on the spacecraft. Later, they will take it to the moon and back. These rockets are attached to the part of the spacecraft called the modules (MOJ-oolz)—the command module, the service module, and the lunar module.

IF YOU PUT YOUR SUPPER DISH TO YOUR EAR, YOU CAN HEAR THE SPACE CRAFT TAKE OFF...

The first-stage Apollo rocket motor is as powerful as 640,000 car engines!

Apollo 15 lift-off

What is a command module?

A command module is the front end of a spacecraft where the astronauts live and do their work. This module is like the cockpit of an airplane and is sometimes called a space capsule. The command module of a spacecraft that goes to the moon has more than two million working parts. An automobile has fewer than two thousand.

What is a service module?

A service module is the part of a spacecraft that carries batteries for electric power. This power is needed for air conditioning, heating, and lighting. The service module also has tanks of oxygen for the astronauts to breathe.

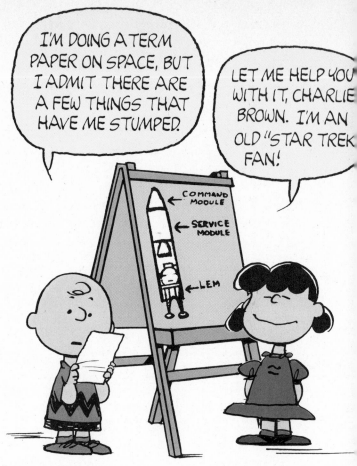

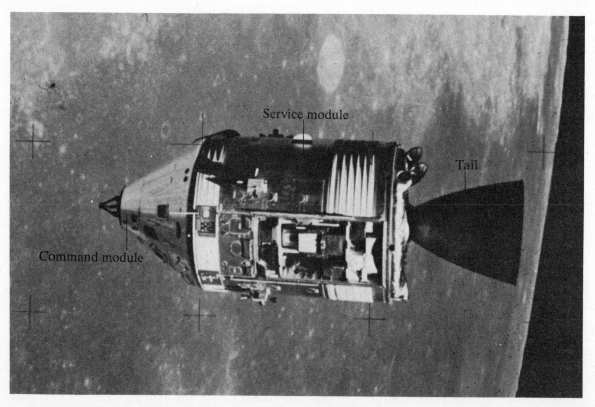

Service module

Tail

Command module

What is a lunar module?

A lunar module is the part of the spacecraft that actually lands astronauts on the moon. It is usually called the LEM, and it is carried inside the third stage of the spacecraft.

After the spacecraft goes into orbit around the moon, the LEM separates from the spacecraft and starts a downward trip. Not all the astronauts go along. One must stay behind to run the orbiting spacecraft.

When the LEM reaches the landing place, its rockets are fired to let it settle down gently on the moon.

The LEM on the moon

113

What happens to the parts of a spacecraft that are dropped off?

As a spacecraft goes up, man-made objects such as rocket stages are left behind in space. The first ones dropped are slowed down by the air. Some of these burn up as they fall. Others splash down into the ocean.

The pieces that are let loose high up go into orbit around the earth. These are known as space junk. They may last for a year or longer before they drop down and burn up. There are more than 3,000 pieces of space junk still in orbit.

A space walk

During a space walk in 1969, astronaut Michael Collins happened to let go of his camera. It became an expensive piece of space junk!

How is a spacecraft steered?

This is usually done by turning the main rocket motors that are at the bottom of the spacecraft. To turn the spacecraft just a little, special small rockets on the sides are fired. All of these steering rockets can be worked by the astronauts or by radio signals from the ground.

What is "mission control"?

All space flights are run from a center called mission control. The people in charge of the flight work at this center.

Mission control does not have to be at the spaceport. It can be hundreds of miles away. The people at mission control talk with the astronauts by two-way radio. This means that the astronauts and the people at mission control have radios that do two things—send out messages and pick up messages. The mission control people watch signal lights and special TV screens to keep track of the flight.

What is NASA?

NASA stands for the National Aeronautics (air-uh-NAW-ticks) and Space Administration. This organization is the part of the United States government that is in charge of exploring space. Thousands of scientists and engineers work for NASA.

How does NASA keep track of a traveling spacecraft?

Radio signals from the spacecraft are picked up by stations on earth. These stations are at several places around the world. Some of the receivers are on ships at sea.

The signals are sent into a computer. It figures out where the spacecraft is at any time.

Can spacecraft be sent to the planets?

Yes. In 1976, an American spacecraft called the Viking landed on the planet Mars. Another spacecraft, Pioneer-Saturn, is now on the way to Saturn and should arrive there in 1979. It has already flown past Jupiter and sent back scientific information. Later, the United States will send another Pioneer to Venus to find out more about this planet. The Russians have already sent two spacecraft, Venera 3 and Venera 4, to Venus. But both crash-landed.

Spacecraft do not have to land on a planet to be useful. They can carry telescopes that take pictures as they pass close to a planet. These pictures are very clear and much better than any taken by telescopes on earth.

Will people ever be able to visit other planets?

Yes, but it will not be easy. A trip will take many months—even to one of the nearest planets, Venus or Mars. The chances are better for a visit to Mars than for a visit to Venus. Although Mars is much colder than the earth, a spacesuit can probably keep people warm enough. But Venus is hotter than the inside of a furnace. A spacesuit would not help there!

What is gravity?

Gravity is a force that every planet, star, and moon has. This force causes everything on or near the planet, star, or moon to be pulled downward.

The pull of the earth's gravity holds the moon in its orbit. The pull of the sun's gravity keeps the planets in their orbits.

Stand on a bathroom scale. Suppose it shows that you weigh 90 pounds (41 kilograms). This means that the downward pull of the earth's gravity on your body measures 90 pounds (41 kilograms).

What is a satellite?

Anything in space that moves in an orbit is called a satellite (SAT-uh-lite). The earth is a satellite of the sun. So are the other planets in our solar system because they orbit the sun. The moon is a satellite of the earth because it orbits the earth. Seven of the planets have satellites moving around them. The earth has only one moon, but Saturn has ten. Jupiter has thirteen! And more moons may be discovered. All these are called natural satellites.

There are also artificial (ahr-tuh-FISH-ull) satellites. The word "artificial" means man-made. Artificial satellites are built on earth and put into orbit. Since the beginning of the space age, hundreds of artificial satellites have been sent into space—weather satellites, TV satellites, and many other kinds.

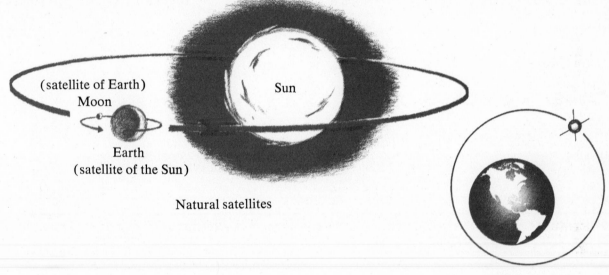

Natural satellites

Artificial satellite

119

What was the first satellite to orbit the earth?

The moon, of course! It is the earth's natural satellite and has been in orbit for billions of years. The first artificial earth satellite, called Sputnik I, was sent up by Russia in 1957. A few months later, the first American satellite, Explorer I, was sent into orbit. After a while, both of these satellites slowed up and dropped closer to the earth. As they fell through the air, they became so hot that they burned up.

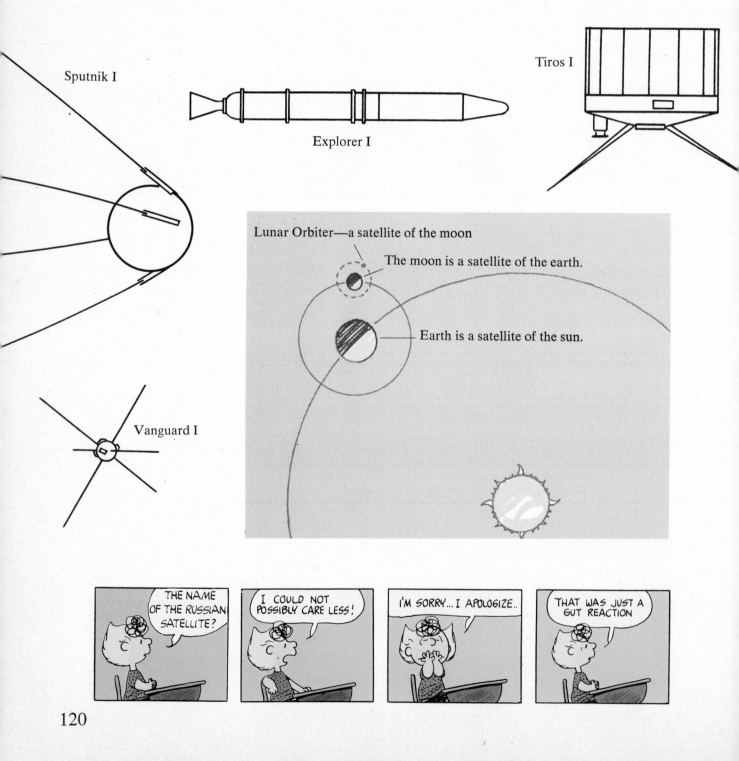

Sputnik I

Explorer I

Tiros I

Vanguard I

Lunar Orbiter—a satellite of the moon

The moon is a satellite of the earth.

Earth is a satellite of the sun.

THE NAME OF THE RUSSIAN SATELLITE?

I COULD NOT POSSIBLY CARE LESS!

I'M SORRY... I APOLOGIZE..

THAT WAS JUST A GUT REACTION

What do artificial satellites do?

Weather satellites orbit the earth several hundred miles up. They measure the temperature and amount of moisture, or dampness, in the air. They send back TV pictures showing where there are clouds and storms on earth.

Communications satellites pick up electrical waves from TV stations. The waves bounce back to distant places on earth. That is how you get "live" TV broadcasts from halfway around the world. Some communications satellites are used for sending long-distance telephone calls.

Some scientific satellites measure radiations from outer space that do not get through the air to the ground. Other scientific satellites carry telescopes that send back pictures of planets and stars. These pictures are much clearer than any taken from earth.

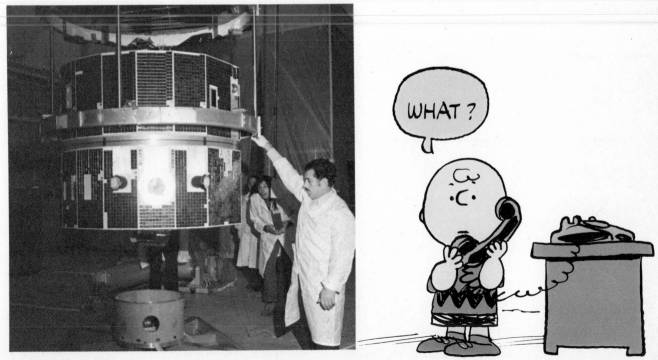

A scientific satellite

121

What is a cosmonaut?

A cosmonaut is a Russian space traveler. The word comes from two Greek words that mean "sailor of the universe." An American space traveler is called an astronaut. The word means "sailor among the stars."

How can you become an astronaut?

If you want to become an astronaut, you must be less than 34 years old, intelligent, and in perfect health. You must have a good education and go through a long testing and training period.

Astronauts must study science and engineering and have at least 1,000 hours of experience flying jet airplanes.

The special training that astronauts get is very difficult. Many are forced to drop out before finishing.

Has a woman ever traveled in space?

Yes. A woman named Valentina Tereshkova (tay-RESH-koe-vah) was one of the cosmonauts in the Russian spacecraft Vostok 6 that orbited the earth in 1963.

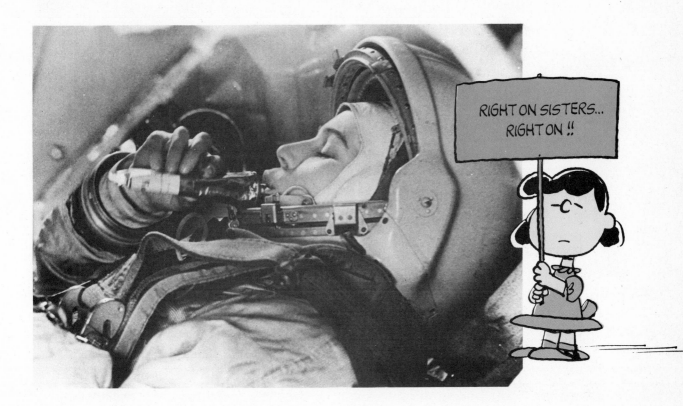

Will there be other women in space?

Probably. In the 1980s, the United States plans to send people into space and back fairly often. NASA is already training a group of air force nurses. They will live and work for a while in space stations that will orbit the earth. These stations will be places for space research. They will also serve as hotels for travelers going from one part of space to another.

Why do astronauts wear spacesuits?

The main purpose of spacesuits is to keep astronauts healthy and comfortable when they are not inside the spacecraft. Each suit is airtight. It keeps the air, the temperature, and the pressure inside the suit as earthlike as possible. The astronauts also wear helmets that have a gold coating on the front. This protects them from the rays of the sun.

Can astronauts take off their spacesuits during a trip?

Yes, if the astronauts stay inside the spacecraft. Of course, they must be "suited up" again when they get ready to take a space walk or land on the moon. They help each other put on the bulky suit. If they are going far from the spaceship, they also hook up a backpack. The pack holds air for breathing, heating and cooling equipment, and radio equipment for contact with earth and with other astronauts.

What is a space walk?

When astronauts go outside their orbiting spacecraft, we say they are taking a space walk. Of course, they are not really walking at all. They are only drifting alongside the spacecraft. Each drifting astronaut is connected to the spacecraft by a long hose. The hose keeps the astronaut from floating off into space. It also takes the place of a backpack. The hose has electric lines for air conditioning and radio. It also has an air tube for breathing. To get back to the spacecraft, the astronauts slowly pull themselves along their hose.

What is a life-support system?

A life-support system has everything that astronauts need in order to stay alive in space. It is the part of a spacecraft that keeps conditions very much as they are on earth.

A life-support system has water, air, and food for the astronauts. It keeps the temperature comfortable. It also protects the astronauts from harmful radiations in space.

The astronauts' backpacks are smaller life-support systems. Astronauts use these outside the spacecraft.

Why do things float around in a spacecraft?

On earth, gravity holds everything down. While a spacecraft is orbiting, the earth's gravity is still pulling on everything in the craft. But another force, which comes from orbiting, also pulls on everything. The two forces are equal and cause everything—and everyone—to float. This kind of floating is called weightlessness. All things inside a spacecraft float around if they are not held down.

FLOATING AROUND IN SPACE CAN BE VERY EXCITING... AFTER A FEW YEARS, HOWEVER, THE EXCITEMENT WEARS OFF!

What do astronauts eat?

Freeze-dried foods are used to save space and to keep things fresh. The food is first frozen, and the ice that forms is then taken out. The astronauts just add water to freeze-dried food, and it is ready to eat.

In an orbiting spacecraft, eating is tricky because of weightlessness. Astronauts cannot drink from an open cup. The liquid forms blobs that float around and wet anything they hit. So drinks must be kept in closed plastic bags. Astronauts must squeeze the liquid right into their mouth. Solid foods are in bite-sized pieces so that crumbs will not float around and pollute the air in the spacecraft.

For long trips, astronauts may someday grow their own food plants in the spacecraft.

How do astronauts get rid of body wastes?

Liquid waste is pumped into space, where it becomes a gas. Solid waste is put into plastic bags with chemicals that kill germs. The bags are thrown away when the spacecraft returns to earth.

How are astronauts made ready for space travel?

Scientists have set up labs on earth that copy the way astronauts will live when they are in space. For example, an astronaut is put inside a large metal ball. The ball is spun in order to put a great push on the astronaut's body. It is like the push people feel when a spacecraft zooms upward. Astronauts also move around under water in spacesuits. That helps them get used to the feel of floating weightless in space.

Also, astronauts work in an exact copy of a command module. They practice using the dials and switches that will control the spacecraft.

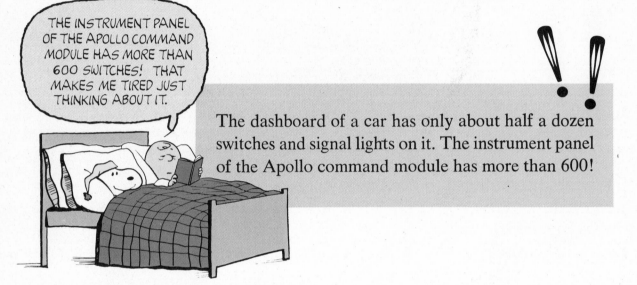

THE INSTRUMENT PANEL OF THE APOLLO COMMAND MODULE HAS MORE THAN 600 SWITCHES! THAT MAKES ME TIRED JUST THINKING ABOUT IT.

The dashboard of a car has only about half a dozen switches and signal lights on it. The instrument panel of the Apollo command module has more than 600!

What is space medicine?

This is the science that takes care of the health of astronauts. Much can be found out in advance about how space will affect the health of space travelers. Doctors study the people being trained in labs on earth. Doctors also check the health of astronauts when they are in space and when they return to earth.

How do doctors check astronauts in space?

Electrical machines connected to the body of an astronaut check his breathing, heartbeat, and temperature. Readings are automatically sent back by radio to doctors on earth.

What does space travel do to an astronaut's thoughts and feelings?

When an astronaut is alone in space for many, many days, he may become very upset. He might even panic. Sometimes things look blurry to him. He feels strange, and things around him do not seem real.

Even if an astronaut is not alone, he may get very tense and grouchy. For this reason, the people who will someday work in satellites called space stations will have to be sent back to earth after about a month. Other astronauts will be sent up to take their places.

MY SISTER CAN BE TENSE AND GROUCHY WITHOUT EVER LEAVING THE BACK YARD.

GET OUT OF MY CHAIR YOU STUPID BIRD. I'M SICK OF WATCHING YOU AND THAT DUMB DOG PLAYING ASTRONAUT!

I HATE BEING MOCKED BY LOW TYPES.

Why is an astronaut sometimes strapped to his couch?

This is done only during lift-off and return to earth. At those times the astronaut's body feels a great push. It is the same kind of push that you feel when you ride in a car that makes a sudden, fast start. It seems as if you are being shoved back into your seat.

The pushing forces are much, much stronger in a spacecraft that is leaving or coming back to earth. The astronaut feels a force of nearly a ton on his body! That is why he must be supported by a couch during take-off and landing.

How long can people stay in space?

Scientists do not yet know for sure. In 1973, a spacecraft named the Skylab was sent into orbit around the earth. There were four men in it. They lived and worked there for almost a month and were in good health when they returned to earth. Other astronaut teams have spent nearly three months in orbit.

There will be many more problems in sending astronauts farther out into space. A trip to a planet will take many months, or even years.

A trip to the nearest star would take a whole lifetime, even if the spacecraft could travel 100,000 miles (160,000 kilometers) a second!

What does the "docking" of two spacecraft mean?

Two spacecraft in orbit can meet and link together. They use their small rocket motors to line up. Then they slowly move toward each other until they can lock together. This is called docking.

In 1975, an American Apollo docked with a Russian Soyuz (SOY-use) in orbit 138 miles (221 kilometers) above the earth. Then the astronauts and the cosmonauts visited back and forth between the two spacecraft.

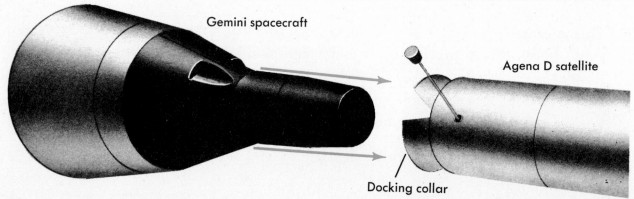

Gemini spacecraft

Agena D satellite

Docking collar

How long does a spacecraft take to go to the moon and back?

The first manned flight to the moon took about four days from the time the spacecraft left the earth until it went into orbit around the moon. The return trip took a little less than three days.

Who was the first person to step onto the moon?

The first person to step onto the moon was Astronaut Neil Armstrong, when he climbed down from the Apollo 11 landing craft on July 20, 1969. Since that time, ten men have walked on the moon.

Neil Armstrong

What does the moon look like close up?

The moon is very rough and rocky. There are tall mountains, deep cracks, and steep cliffs. All over the moon are thousands of scooped-out holes shaped like saucers. These holes are called craters. Many of them were formed when rocks, flying through space, crashed onto the moon. Most of these rocks exploded when they hit the moon. Bits of them were scattered all over. So you cannot find the rocks, but you can see the holes they made. Some craters are less than a foot (30 centimeters) wide. The biggest ones are more than 150 miles (240 kilometers) wide.

There are some large, smooth places on the moon called "seas." These have no water in them, just rocks and soil. The seas were formed by melted rock that spread out, cooled, and became hard. The melted rock may have come up from the hot inside part of the moon. Or large rocks that crashed onto the moon may have melted from the heat of the crash.

Moon rover

Two million football stadiums could fit inside the largest moon crater! **!**

132

How do astronauts talk to each other on the moon?

They use the small radio that is built into each spacesuit. Radios work on the moon because radio waves can travel even where there is no air.

As astronauts talk to each other, mission control listens in. The people on earth answer questions and tell the astronauts what to do.

Could an astronaut hitchhike on the moon?

Yes, if another astronaut happened to be coming along in a moon rover. The rover looks like a jeep or a dune buggy. It gets its power from batteries. On one moon expedition, astronauts David Scott and James Irwin traveled more than 17 miles (27 kilometers) in their moon rover. They collected moon rocks to take back to earth.

Is there gravity on the moon?

Yes. But the moon is much smaller than the earth, so its gravity is much weaker. If you weighed 90 pounds (41 kilograms) here on earth, you would weigh only about 15 pounds (7 kilograms) on the moon.

Why do astronauts shuffle along instead of walk on the moon?

Because the moon's gravity does not pull as strongly as the earth's gravity, astronauts cannot walk on the moon the same way they walk on earth. If they did, they would rise a few feet off the ground with every step. They can keep better control and stay on the ground by just shuffling along. If astronauts did not have to wear their heavy spacesuit and backpack, they could jump 35 feet (11 meters) high.

Apollo-Saturn 1 at Cape Canaveral

Gemini VII spacecraft

Astronaut Edwin Aldrin on the moon

David R. Scott and Apollo 9

Athletes on the moon could leap over a two-story house. And they would come down no harder than they had after a 6-foot (180-centimeter) jump on earth!

WOW, EVEN A LITTLE GUY LIKE ME COULD BREAK ALL HIGH JUMPING RECORDS... ON THE MOON, THAT IS.

How do astronauts leave the moon?

When astronauts finish their moon work, they blast off in the LEM. It goes up to meet the main spacecraft that is orbiting the moon. The two ships dock together and the astronauts board the command module. The LEM is left behind as space junk. The third-stage rockets on the service module are then fired, and the spacecraft heads back toward earth.

How a spacecraft gets out of orbit.

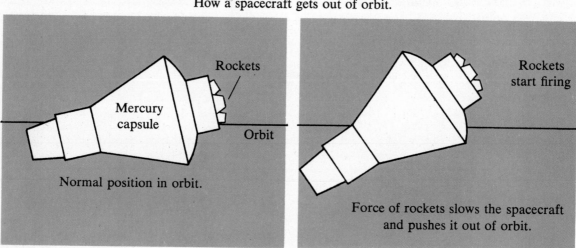

Rockets

Mercury capsule

Orbit

Normal position in orbit.

Rockets start firing

Force of rockets slows the spacecraft and pushes it out of orbit.

What is re-entry?

As a spacecraft returns from outer space, it must plunge into the air before it can land. This is called re-entry.

136

What is the heat shield on a spacecraft?

When a returning spacecraft plunges back into the earth's air, its gets extremely hot. To protect the astronauts, the front end of the capsule is covered with a heat shield made of special plastic. The shield heats up to about 5,000 degrees Fahrenheit (5,000° F., or 2,700° C.). Some of the plastic melts and burns off, taking away the dangerous heat. Inside the capsule, the temperature stays at a comfortable 80 degrees Fahrenheit (27° C.).

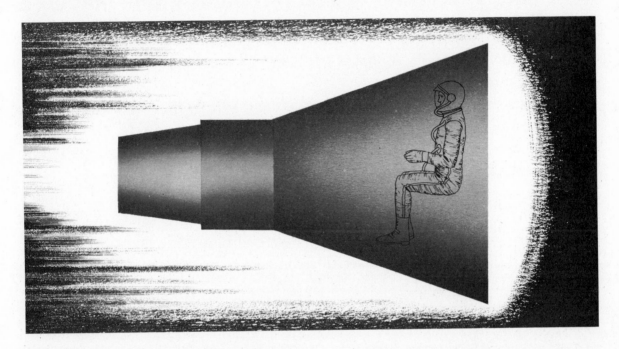

Can the astronauts talk with mission control during landing?

Astronauts and mission control talk back and forth by radio right up to the time the spacecraft comes back into the air. Then, as the heat shield begins to get hot, a strange thing happens. A cloud of tiny electrical bits gathers around the command module. Radio waves cannot get through this cloud. So, for several minutes, there is only silence between the astronauts and the ground.

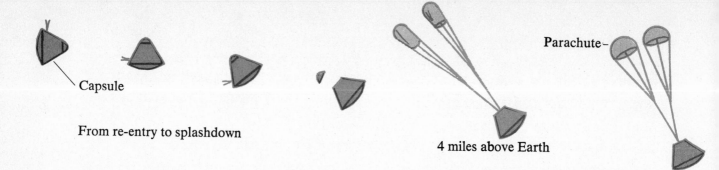

Capsule

From re-entry to splashdown

4 miles above Earth

Parachute

How does a returning spacecraft make a safe landing?

As the spacecraft gets ready to enter the air again, the service module is thrown off and left in space. Now all that is left is the command module—the capsule —with the astronauts in it.

Before re-entry, the capsule is turned around so that the end with the heat shield faces forward. This is done by firing small steering rockets. As the capsule enters the air, it does not point straight toward the ground, it comes in on a slanting path.

Now a computer takes over the steering. If the capsule were aimed too low, it might fall downward, get too hot, and burn up. If it were aimed too high, it might bounce off the upper layers of the air and be thrown back into space. Because it no longer has its powerful main rocket motor, the capsule would not have the force behind it to return to earth.

At the time of re-entry, the capsule is moving at 25,000 miles (40,000 kilometers) an hour. For a safe landing, this enormous speed must be cut down to only a few miles an hour. Rubbing against the air makes the capsule lose most of its speed. Then, about 4 miles (6 kilometers) above the earth, two small parachutes open. They slow the falling motion even more. They also keep the capsule from wobbling. About 2 miles (3 kilometers) up, three big parachutes open. The capsule floats down to earth at a safe speed.

What is "splashdown"?

This is the moment a capsule lands in the earth's water. As soon as splashdown takes place, ships and helicopters rush to the floating capsule. Divers jump into the water and place a doughnut-shaped balloon around the capsule to make sure it does not sink. The astronauts open a door called a hatch, and are lifted into a helicopter. They are taken to a nearby ship. Their space trip is over.

Up to now, all American spacecraft landings have been made on the ocean. The Russian cosmonauts bring their spacecraft down on land.

The Apollo capsule was tested for splashdown by dropping it into a tank of water from a tower 18 stories high!

2 miles above Earth

What is a space station?

It is a special kind of satellite that will circle the earth a few hundred miles up. One kind of space station will be shaped like a huge ring, hundreds of feet across. It will be put together in orbit after the parts are sent up from the ground.

The station will be turning as it moves along its orbit around the earth. The turning motion will create a kind of gravity. It will allow people to walk around as on earth, instead of being weightless and floating.

What will a space station be used for?

A space station will be a place where scientists can work in space. It will also be a stopping-off place for spacecraft that are going farther out into space. The inside of the space station will be divided into science workrooms and labs. There will also be dormitories, kitchens, and even a gym.

Will there be factories in space?

Probably. They will be useful in making things that are hard to make on earth. For example, metals can be joined together by heating them. This is called welding. If a weld is to be strong, the pieces must not be touched by air during the joining. On earth, small objects can be welded in a closed box with the air pumped out. Many things are too big to be welded this way. But welding would be easy in a factory where there is no air—on the moon or on a space station orbiting the earth.

What is a space shuttle?

A space shuttle is a spacecraft that will travel back and forth between the ground and an orbiting space station. It will carry people and heavy loads of material. It will have wings and be about as big as a jet plane.

As a shuttle heads upward, some of its rockets will drop back to earth by parachute. These rockets will be used again and again.

When starting its return trip, the shuttle will fire a rocket motor and zoom safely through the air, protected by a heat shield. Then the shuttle will land on a runway, as an ordinary airplane does.

An airplane trip in the United States costs about 20 cents a mile (12 cents a kilometer). A trip on the shuttle may cost 10 dollars a mile (6 dollars a kilometer). If so, a trip to the moon and back would cost more than 4 million dollars!

OF COURSE, YOU COULD SAVE TWO MILLION WITH A ONE WAY TICKET!

What is a space colony?

A space colony is a kind of island in space where thousands of people can live and work. No space colony exists yet, but the United States and Russia are both planning colonies.

Each space colony will be a huge aluminum tube half a mile (nearly a kilometer) long, orbiting the earth. The tube will keep turning slowly to give the feeling of gravity. Large mirrors will focus the sun's rays and provide all the power for electricity needed by the colony. Once the colony is set up, people will not need to bring many new things from the earth. They will raise their own food crops and farm animals. They will be able to get minerals and other building materials from mines on the moon. So people in space colonies will be helping the earth save food, minerals, and other natural resources.

Why should we build space colonies?

In a space colony, there will be plenty of food. There will be no earthquakes, no floods, and no storms. So a space colony will be a safer place to live than earth is.

A space colony will have sunlight 24 hours a day. The sun's rays will give the colony all the power it needs to run its machines. If the world ever becomes too crowded, people can move to space colonies.

! You and your family could be living in a space colony by the time you reach middle age! !

Are there people living somewhere else in space?

Scientists do not know of any place in the universe besides the earth where there is life. But hundreds of billions of stars are in the universe. Millions of these stars may have planets orbiting around them. It is very likely that on many of those planets conditions are right for some kind of life.

There could be intelligent creatures somewhere out there. They might be trying to get in touch with us by sending out radio signals. Scientists have searched for such signals, but so far they have not found any.

Index

References to pictures are in *italic type*.